"Urban studies has been heavily shaped over the past half century by serial bursts of scholarship focusing on a small number of cities, or parts of cities. Through this process certain cities or urban districts have emerged as exemplars of urban change everywhere to such an extent that they can even emerge as clichés. In this topical and innovative collection, authors and contributors confront the phenomenon of over-researched places head on. Drawing on a range of lucid case studies from across world the book's contributors confront for the first time what it means for urban studies disciplines to be so heavily reliant on a small number of paradigmatic cases."

Stephen Graham, *Newcastle University, UK*

"Button and Taylor Aiken are to be applauded for bringing together this brilliant, long overdue and incisive examination of the concept and processes of over research. Globally framed and theoretically and empirically nuanced, each of the chapters push us to rethink what over research means and reconsider its implications for social research."

Sarah Neal, *University of Sheffield, UK*

Over Researched Places

The book explores the implications that research-density has on the people and places researched, on the researchers, on the data collected and knowledge produced, and on the theories that [illegible]re developed.

It examines the effects that [illegible]h-density has on the people and places researched, on the researche[illegible] collected and knowledge produced, and on the theories that ar[illegible]ing together experiences from a variety of countries and [illegible]ies and research methods, the volume outlines the ro[illegible]es from and what can be done about it.

The book will [illegible]archers working in ethnographic disc[illegible]logy, Urban Planning, and Sociology [illegible]esent' of already existing research on [illegible]

Cat Button is a senior lecturer in [illegible]g and Landscape at Newcastle University, [illegible]ary and international research on the global challe[illegible]urrently a co-investigator in two UKRI GCRF Hubs: *W*[illegible] *Sustainable Development* and *Living Deltas*.

Gerald Taylor Aiken is a research associate at the Luxembourg Institute of Socio-Economic Research (LISER) and a fellow at the Rachel Carson Center for Environment and Society. He researches the role of community in pursuing low-carbon futures, particularly how community is used to understand, value, and relate to the environment.

Routledge Studies in Human Geography

This series provides a forum for innovative, vibrant, and critical debate within Human Geography. Titles will reflect the wealth of research which is taking place in this diverse and ever-expanding field. Contributions will be drawn from the main sub-disciplines and from innovative areas of work which have no particular sub-disciplinary allegiances.

Why Guattari? A Liberation of Cartographies, Ecologies and Politics
Edited by Thomas Jellis, Joe Gerlach, and John-David Dewsbury

Object-Oriented Cartography
Maps as Things
Tania Rossetto

Human Geography and Professional Mobility
International Experiences, Critical Reflections, Practical Insights
Weronika A. Kusek and Nicholas Wise

Geographies of the Internet
Edited by Barney Warf

Locating Value
Theory, Application and Critique
Edited by Samantha Saville and Gareth Hoskins

Spatialized Islamophobia
Kawtar Najib

Over Researched Places
Towards a Critical and Reflexive Approach
Edited by Cat Button and Gerald Taylor Aitken

For more information about this series, please visit: www.routledge.com/Routledge-Studies-in-Human-Geography/book-series/SE0514

Over Researched Places

Towards a Critical and Reflexive Approach

Edited by Cat Button and Gerald Taylor Aiken

Routledge
Taylor & Francis Group
LONDON AND NEW YORK

First published 2022
by Routledge
4 Park Square, Milton Park, Abingdon, Oxon OX14 4RN

and by Routledge
605 Third Avenue, New York, NY 10158

Routledge is an impint of the Taylor & Francis Group, an informa business

British Library Cataloguing-in-Publication Data
A catalogue record for this book is available from the British Library

Library of Congress Cataloging-in-Publication Data
Names: Button, Cat, editor. | Taylor Aiken, Gerald, editor.
Title: Over researched places : towards a critical and reflexive approach / edited by Cat Button & Gerald Taylor Aiken.
Description: Abingdon, Oxon ; New York, NY : Routledge, 2022. | Series: Routledge studies in human geography | Includes bibliographical references and index.
Identifiers: LCCN 2021052196 (print) | LCCN 2021052197 (ebook) | ISBN 9780367567712 (hardback) | ISBN 9780367567750 (paperback) | ISBN 9781003099291 (ebook)
Subjects: LCSH: Human geography—Research. | Ethnology—Research. | Sociology, Urban—Research. | Interdisciplinary research.
Classification: LCC GF26 .O84 2022 (print) | LCC GF26 (ebook) | DDC 304.2072/1—dc23/eng/20211223
LC record available at https://lccn.loc.gov/2021052196
LC ebook record available at https://lccn.loc.gov/2021052197

ISBN: 978-0-367-56771-2 (hbk)
ISBN: 978-0-367-56775-0 (pbk)
ISBN: 978-1-003-09929-1 (ebk)

DOI: 10.4324/9781003099291

Typeset in Times New Roman
by Apex CoVantage, LLC

Contents

Contributors

Gerald Taylor Aiken is a research associate at the Luxembourg Institute of Socio-Economic Research (LISER) and a fellow at the Rachel Carson Center for Environment and Society. He researches the role of community in pursuing low-carbon futures, particularly how community is used to understand, value, and relate to the environment.

Cyril Blondel is an associate professor of geography at the University of Reims Champagne-Ardenne, France. He teaches political geography, epistemology of social sciences, and urban and regional planning. Mobilising critical theory, his research takes inspiration from decolonial thinking and aims at unpacking European development policy logics towards its own peripheries/borders and marginalised populations.

Cat Button is a senior lecturer in the School of Architecture, Planning and Landscape at Newcastle University, UK. She creates interdisciplinary and international research on the global challenges of water. She is currently a co-investigator in two UKRI GCRF Hubs: *Water Security and Sustainable Development* and *Living Deltas*.

Pratichi Chatterjee is an urban geographer with a PhD (University of Sydney) on contemporary housing and land dispossession and connections with Australia's settler colonial present. Her research interests include geographies of dispossession and critical perspectives on property, race, and colonisation. Pratichi is a researcher of racism and homelessness for crisis in London.

Jenna Condie is a senior lecturer in Digital Society at Western Sydney University. Her interdisciplinary research traverses critical psychology, geography, and technology studies. Her work is orientated towards enabling equitable mobilities, just places, safe spaces, and emancipatory technologies.

Marine Duc is a PhD student at the University of Bordeaux Montaigne. She is currently a research and teaching assistant at Gustave Eiffel University. Her PhD focuses on the making of race through student migrations between Greenland and Denmark and she is interested in power relations in the fields of research and education.

Alejandra de Bárcena Myrsep has a BA in Middle East Studies and International Relations from the University of Exeter where she was inspired by the academic work at the European Center of Palestinian Studies. She has recently graduated with an MA in Applied Cultural Analysis from Lund University.

Hanna A. Ruszczyk is a feminist urban geographer at Department of Geography, Durham University. She has published on the liveability of small cities, hope, aspirations, and resilience of urbanising cities in *Antipode*, *Area*, *Disasters*, *Environment & Urbanization*, and *Urban Geography*. Before academia, Hanna worked for the ILO and UNDP.

Lise Serra (PhD) is a senior lecturer at the University of Reunion Island, France. As a fieldwork scientist, she explores relationships between construction sites and urban projects in urban areas as a way to focus on the making of the contemporary city.

Chandni Singh is a researcher at the Indian Institute for Human Settlements. She examines the drivers of differential vulnerability to climate change, linkages between adaptation and development, and behavioural aspects. She is a lead author of the IPCC Assessment Report 6 and serves on the editorial boards of *Regional Environmental Change*, *Climate and Development*, and *Urbanisation*.

Alistair Sisson is a postdoctoral research fellow of the School of Geography and Sustainable Communities, University of Wollongong. He does research about housing, stigma, gentrification, urban planning, and urban governance.

Laura Wynne is an adjunct fellow with the Institute for Sustainable Futures at the University of Technology Sydney. Her research has focused on sustainable and equitable urban policy, focused particularly on the experiences of public housing tenants in areas targeted for housing renewal.

Marielle Zill is currently a postdoctoral researcher at the Athena Institute at the Vrije Universiteit Amsterdam. The chapter was produced as part of her PhD at Utrecht University, which focuses on the spatial, material, and institutional differences of asylum seeker accommodation and their influence on familiarity and estrangement.

Over-research

What, why, when, where, how?

Cat Button and Gerald Taylor Aiken

What is over-research?

Over-researched places and topics are something that many scholars experience. Most researchers have a story to tell about their experiences of over-research when we introduce this topic. Worryingly, more than a few research participants also nod knowingly when we mention over-research to them. However, very few talk about it freely or write about over-researched places. Over-research seems to be something reflected on when pointed out but not an issue considered in the planning stages of many projects. This volume aims to change that. Social researchers increasingly need to take account of the presence of other researchers, prior and concurrent studies, saturation of data, and the existing and sedimented narratives that can override their research findings. There is a need for reflexive interrogation of researcher saturation and its consequences. Research itself, and theory building more widely, can be weaker where it is over-reliant on examples which may prove to be outliers or when the applicability of generalisations is over-claimed. Over-research also produces a sample bias: familiar cases are easier to communicate to other researchers; possibly easier to publish; or conversely, researchers wring dry popular cases. This also raises questions on the nature of research itself: is it possible to over-research anything or is seeming *over*-research just poor research? There may be a need for more research in over-researched places but on under-researched topics or groups. This book explores the consequences of theory being developed from research on places that are saturated with other researchers from multiple disciplines.

Over-researched places shape the discourse as key case studies in many disciplines. Conducting research in a research(er)-saturated space can make the process easier but can also have drawbacks such as research fatigue of participants. This book builds on the ideas of research fatigue but goes further to consider the wider consequences of over-researched places. Research fatigue is just one ethical and practical concern within over-researched places as explored by Taylor Aiken in Chapter 1 of this volume. There is little written directly on over-research in English, although a special issue of a French-language journal is forthcoming (Chossière et al., 2021). This book is thus based on original thinking and primary research. With the growth of interest in critical approaches to methodological training, a number

DOI: 10.4324/9781003099291-1

of edited volumes and monographs have been published on critically reflective methods in recent years. As readers of these collections ourselves, we have found them extremely helpful. We have found, however, that many of the approaches represented in this literature ended up replicating the very same individualist and technocratic frames of research we seek to critique. Further, despite thorough and embedded methodological training programmes in universities, over-research persists. This collection, featuring chapters from experts engaged in paradigmatic research fields from around the world, aims to fill this gap with empirical and theoretical illustrations of the various means through which over-research comes to be and what can be done to ameliorate it.

Why, despite a long and distinguished history, can methodological training in social research still result in the phenomenon of over-research? Incr`easingly, we, as a community of researchers and practitioners, have come to recognize the barriers and power differentials that prevent more effective research in our various fields. We see over-research is often considered as negative for researchers, those researched, and also for the research itself. This collection is therefore a critical one. We present chapters that help us to analyse the research process in ways that are accessible and productive, offer practitioners entry points into seemingly impenetrable issues, and work to bridge the gaps in current approaches to social research.

Some of the chapters here pay more attention to the effects of over-research on research itself—for instance, Duc's exploration of the role of places in Greenland and how they are represented academically. Others pay close attention to over-research's close relationship with research fatigue and outline an ethical case for paying attention to over-research in various different studies. For example, Sisson et al.'s ethical injunction to see the research process through the eyes of those most affected by it. In addition, as Serra's study on Le Duchère shows, over-research combines the research subjects but also muddies the research process as well.

Why a book on over-researched places?

'Over-researched Places' was a session at the RGS-IBG 2018 Conference, sponsored by the Social and Cultural Geography Research Group, the Political Geography Research Group, and the Participatory Geographies Research Group. It was a lively conference session that sparked discussions in the room and also gained attention on social media. This book puts forward the ideas from the presented papers and questions along with additional complementary chapters across several disciplines.

The initial starting point for these discussions and this book comes from observations and conversations between the editors (Taylor Aiken and Button) who did their PhDs together. While we researched very different contexts (the everyday environmental politics of community in the UK and access to water supplies by the middle-classes in Mumbai), we noted a very similar occurrence and frustration: both our fields were overpopulated in one or two high-profile 'honeypot sites'. While we searched for sound literature on how to manage and deal with this, we mainly coped with reflecting on this between ourselves and innovating bespoke techniques and solutions. We found that there was a wealth of literature on methods and we used some of it and adapted for our needs. However, we have found no systemic, focused attempt to explain, explore, theorise, or understand over-research itself. Hence, we decided to draw together this volume to start this much-needed conversation.

When do we need to consider over-research?

Over-researched places is an idea whose time has come. The discussion and debate need to start now. We want to see how the informal chats we have had with so many colleagues over the past decade or so lead to some serious reflections and changes in approach. The methodological writings on addressing concepts such as research fatigue now need to be drawn together and pulled into conversation with theoretical debates about 'best practice' and the dominance and over-representation in the literature of places, cases, and even certain scholars and studies.

The other key temporal question we raise is when in the research process over-research should be considered. This is a more complex question. We should consider how we choose places as case studies right at the beginning of the process (see Singh in this volume). Reflection after the research encounter proper is also important and we look back at what we have researched to consider the effects that an over-researched place may have on our data and on the ideas we have produced.

Where are over-researched places?

Over-researched places are all over the world. This edited collection draws together international experiences to consider the implications that research-density has on the people and places researched, on the researchers, on the data collected and knowledge produced, and on the theories that are developed. It thus gets at the heart of what it is to do research and produce knowledge and calls for reflexivity in practices. Certain places are magnets for researchers and sometimes we bump into other researchers or share interview appointments with them. The 'Ghosts of Researchers Past' linger at case study sites and its traces are present in the work we produce (as explored further by Button in Chapter 4 of this volume).

The phenomenon of over-researched places cuts across disciplines, starting with those represented in this book (geography, urban planning, housing, anthropology, and cultural studies) but reaching far beyond. There is a nexus of discipline, topic, methods, and location that can be seen to result in over-researched places. So, one location might be over-researched from one angle or using one approach but could be entirely overlooked from another. The flip side is the concept of places being ignored and overlooked and this is something explored within this volume (directly by Ruszczyk and also by Myrsep and in the chapter by Singh).

There exist many harbingers of what we talk about as over-research that are already well-established phenomena in critical approaches to qualitative social research. For instance, research fatigue for participants or the role of lighthouse case studies in particular fields has explored the problems and possible solutions to some aspects of what we outline as over-research. This book particularly investigates the role of *place* as the locus of this phenomenon. The places explored here range from narrow and specific places such as a hotel (Zill in Chapter 9) to a neighbourhood (Sisson *et al.* in Chapter 2; Serra in Chapter 5), a city (Button in Chapter 4; Ruszczyk in Chapter 7), or even a whole region (Blondel in Chapter 3; de Bárcena Myrsep in Chpater 6; Duc in Chapter 8). This combines focus on participants, on research practices, and also on the culture and lived materiality of certain places. This book thus brings together existing threads of critical research in this area and provides a platform for an increasingly relevant and emerging phenomenon: over-research.

How does this book address over-research?

There are three sections in this book to present a systemic and theoretical perspective on what over-research is, how it comes to be, and what can be done about it, although most chapters cut across all three thematic areas. This volume, for the first time, gathers together various reflections and experiences of over-research and looks ahead to what can be done about it. Doing so, it gives the pervasive experience of over-research a name and allows something to be done about this phenomenon. By weaving together experiences from various countries and across disciplinary boundaries, the volume outlines the roots of over-research, where it comes from, and most importantly, what can be done about it. This volume helps to navigate the tricky 'absent presence' or even 'present presence' of already existing research on their fields of exploration.

Our take-home messages from pulling together this collection has two main goals. First, to consider the future direction for critical social research and also to reflect on the potential for linking insights from this body of work with practitioners. On the research front, we would like to identify substantive topical and practical areas that have not yet been given specific and careful treatment. We would also like to propose further developments in theory, thinking about how different theories have been used by our authors, bodies of theory that are currently underutilised, and the potential for further development of specific theories that clearly strike a chord here ('participation' and 'engagement' emerge as two of these). We would also like to consider the implications of awareness of over-research on the training and development methods of emerging scholars. Ideally, future conversations will include tips and techniques for those teaching research projects to include over-research as a separate stand-alone object of attention. In relation to practice, we use this space to offer a critical framework through which practitioners can understand their activities, offering a series of powerful questions with which they can interrogate their work and engagement with researchers. This might include asking questions about subjectivity (who is this research designed for? what do we anticipate that they do within it?) and difference (who is excluded by this research? who stands to gain from it?).

References

de Bárcena Myrsep (2022) 'Research has killed the Israeli-Palestinian conflict': navigating the over-researched field of the West Bank, in: Button, C. and Taylor Aiken, G. (Eds.)

Over-Researched Places: Towards a Critical and Reflexive Approach. Routledge, New York. pp. 81–99.

Blondel, C. (2022) Epistemological, decolonial, and critical reflections in constructing research in former Yugoslavia, in: Button, C. and Taylor Aiken, G. (Eds.) *Over-Researched Places: Towards a Critical and Reflexive Approach.* Routledge, New York. pp. 37–56.

Button, C. (2022) Ghosts of researchers past, present, and future in Mumbai, in: Button, C. and Taylor Aiken, G. (Eds.) *Over-Researched Places: Towards a Critical and Reflexive Approach.* Routledge, New York. pp. 57–69.

Chossière, F., Desvaux, P., Mahoudeau, A. (2021) Les enjeux de la surétude en sciences sociales. Annales de Géographie. 2021/6 (N. 742), p. 140.

Duc, M. (2022) When over-researchedness is invisibilised in bibliographic databases: insights from a case study about the Arctic region, in: Button, C. and Taylor Aiken, G. (Eds.) *Over-Researched Places: Towards a Critical and Reflexive Approach.* Routledge, New York. pp. 111–132.

Ruszczyk, H. (2022) Overlooked cities and under-researched Bharatpur, Nepal, in: Button, C. and Taylor Aiken, G. (Eds.) *Over-Researched Places: Towards a Critical and Reflexive Approach.* Routledge, New York. pp. 100–110.

Serra, L. (2022) La Duchère, Lyon, France: an over-researched place that ignores itself, in: Button, C. and Taylor Aiken, G. (Eds.) *Over-Researched Places: Towards a Critical and Reflexive Approach.* Routledge, New York. pp. 70–80.

Singh, C. (2022) Locating climate change research: the privileges and pitfalls of choosing over-and under-researched places, in: Button, C. and Taylor Aiken, G. (Eds.) *Over-Researched Places: Towards a Critical and Reflexive Approach.* Routledge, New York. pp. 149–166.

Sisson, A., Condie, J., Chatterjee, P., and Wynne, L. (2022) Overcoming over-research? Reflections from Sydney's 'Petri dish', in: Button, C. and Taylor Aiken, G. (Eds.) *Over-Researched Places: Towards a Critical and Reflexive Approach.* Routledge, New York. pp. 23–36.

Taylor Aiken, G. (2022) Towards a theory of over-researched places, in: Button, C. and Taylor Aiken, G. (Eds.) Over-Researched Places: *Towards a Critical and Reflexive Approach.* Routledge, New York. pp. 5–22.

Zill, M. (2022) Confessions of an 'academic tourist': reflections on accessibility, trust, and research ethics in the 'Grandhotel Cosmopolis', in: Button, C. and Taylor Aiken, G. (Eds.) *Over-Researched Places: Towards a Critical and Reflexive Approach.* Routledge, New York. pp. 133–148.

1 Towards a theory of over-researched places

Gerald Taylor Aiken

Introduction

What is over-research and what can we do about it? This chapter sets over-research in the context of wider trends and patterns in academic research. Framing over-research as 'over' points out that totemic case study locations in any topic—which may or may not be fair representatives of the field—are a problem. First, and most basically, it is a problem in social research for those being researched. When one group or place becomes 'The Place' where this research is done, it seems that everyone wants a piece of the action and those on the ground can become swamped. Research has developed tools and techniques to deal with this from participative action research, coproduction of knowledge, or more light touch, remote-sensing qualitative methodologies. However, the argument here is that while these techniques can come to terms with research fatigue, they *do not address over-research in a more systematic way.*

It is also unfair to expect research subjects to be organised enough to adequately hold off pushy researchers, especially research on/with more marginalised people and places without the resources to cope with or the awareness of what is at stake. These issues, which often go under the term *research fatigue*, are a problem not only for the research participants but also for the research itself. In qualitative research, interviewing and ethnographic and participative evidence and experience gathering are dulled by being 'yet another' researcher among a steady stream. As the quantity of research increases, interviewees can become jaded. Nevertheless, over-research is a more encompassing phenomenon than research fatigue. Over-research can also fail to accurately represent the issues at stake and can also worsen the quality of the data gathered. Representationally, it can also over-simplify complex phenomena. The chapter argues that over-research is a structural issue—or at least a deeply sedimented one—in current academic research. Over-research arises as much from a researcher's social context as their individual choices. That is, the way researchers are socialised plays as much a role in producing over-research as an individual choice or decision. Importantly, this lessens the moral responsibility of researchers. It also allows an analysis to be made of under-research as a corollary to, and a coexisting challenge to, over-research.

Thus, this chapter first addresses over-research, what it means, and how it comes to be. In particular, the chapter argues there is a lot to be gained by separating

DOI: 10.4324/9781003099291-2

over-research from research fatigue. These two often come together in the literature but I want to argue that they are significantly distinct. By surveying noted effects such as bandwagoning, the Matthew effect, and Price's law, this chapter tries to offer some theoretical diagnostic tools that help understand what over-research is and thus begin to chart a way to think through what can be done about it, distinct to research fatigue. The chapter then ends with an example of a deeply sedimented case of over-research: Chicago. This is thus a more abstract and analytical chapter than the rest of this book which is more empirically grounded, hopefully helping to produce a more balanced and integrative collection.

Beginning to research over-research

One topic of rapidly expanding research interest, verging on over-research, is the Transition Town movement. *Transition* is a model of community-based DIY environmentalism that, in the decade since its founding in 2007, has become the go-to example of a range of community initiatives for sustainability, at least in the UK. Much academic attention also followed. As one of the first PhD students studying this emerging Transition Town movement, I had a box-office seat to witness this rapid expansion. When I shifted from volunteering, and being involved in environmental activism as a participant, to being paid to study the same movements (very fortunately with a scholarship to carry out a PhD), I set up search alerts, Google Scholar notifications, and RSS feeds to let me know whenever any new relevant information was published. Early on, I would eagerly lap up these alerts and anticipate any new article being published on the Transition's model of community-based environmentalism—not only academic articles but also blog posts, recorded lectures, or online published Master's theses. Very soon, however, it became impossible to keep up with everything that was published on this topic. Transition rapidly expanded as an area of interest and the sheer quantity of literature being produced became overwhelming. I had to learn to be selective. There is no way any one researcher could have read everything on even this very specific empirical example of Transition Towns. Eventually, I would skip the outputs that looked less appealing, until it got to the point where I was able to read less than half of what was published. Eventually, I was only reading, and then just scanning, a fraction of even what my alerts informed me of. When the time came to shift jobs and e-mail addresses, I did not set up these alerts again. No doubt I missed lots of excellent insights, but even sifting through these alerts had become a significant task. So, when thinking through reasons why new researchers can be covering the ground already researched before—even asking the same questions to the same people—I want to hold the individual responsibilities of the researcher lightly. I know how easy it can be to skip research already carried out—even very good research on a highly relevant and specific topic. I was embedded in the field of human geography and, rather sheepishly and embarrassingly, would note years later that something highly relevant to my own research was already published in sociology, anthropology, or political science. No doubt, I missed much more. Given this experience, I want to marry an individual or case-by-case analysis to a more diagnostic one—to hold the whole manner in which academia burgeons in productivity, outputs, and publications responsible too. Together with the ways in which we as researchers have been enjoined to turn every research observation into an output, attempting to understand what was going on here led to more readings on over-research itself, conversations with colleagues at the time (shout out to Cat Button), and, eventually, this volume. This section reviews some of the most helpful writings, but also, given this experience, tries to push them for a more comprehensive understanding and diagnosis of over-research.

There is a small but growing selection of writings outlining that over-research exists, defining it, showing that it is a problem, and even some tentative suggestions as to what can be done about it. What many of these pieces miss though is a direct examination of over-research itself, and more helpfully, an outline of why. *What is it exactly that produces over-research? By what process and mechanisms does research tend towards this 'bunching up'?* The answer to this needs to hold in tension both the reasons why individual researchers choose to be drawn towards the same honeypot sites, alongside a more structural critique that analyses the underlying conditions of why research itself tends towards the 'usual suspects'. Going further, helpful analysis would ask how those usual suspects come to be produced in the first place. To be clear, not every study or subfield has the same gravitational pull towards the same site or topic but there is, at the very least, a pervasive over-representation in research: whether an iconic study, a 'rock star researcher', a famous case study location, or a social process that is seen as more intuitive or graspable than what it purports to represent. Building on perhaps the signature publication in this sub-field, Neal et al. connect the over-researched places to what Gilroy (1987) called the 'symbolic location' of places. In this case of over-researched places, certain places, events, or practices have this symbolic allure or charm. In this section, to start the chapter, I take three of the most clear and insightful articles on over-research—outlining what they add and how they define the discussion of over-research, before pushing forward and grappling with the missing why. I argue that this comes from analysing over-research separately from research fatigue.

Clark (2008) addresses researchers' perspectives of over-research. Sukarieh and Tannock (2013) examine over-research from the perspective of those being researched. Meanwhile, Neal et al.'s (2016) discussions of Hackney outline the specific place that can be over-researched. Together, these three represent the state of the art on over-research but there are still many unexplained aspects: the relationship between over-research and research fatigue or allocating to what extent over-research is a failure of researchers themselves or research, in general. That is, how individual or structural a phenomenon is over-research? Or, to use more phenomenological language, to what extent is over-research a result of the 'thrownness' of current academic political economy or can it be seen as a voluntary action?

Clark (2008)

Over-research literature often outlines personal tips for researchers so that they can fairly manage unequal relationships of power with those researched: specifically, between the researcher and those researched. Researchers can be conscious of how much of interviewees' time they take up and how much more 'extraction' takes place when rephrasing and representing someone's lifeworld into academic-ese. More rarely, authors like Clark (2008) outline the effects of research fatigue and over-research *on the research process*.

Social research that seeks to describe and explain the world inevitably brings researchers into contact with people. Without these people—'research subjects'—social and qualitative research would be impossible. Therefore, getting the right relationship between researchers and research subjects is a fundamental part of research. While many resources outline this from the perspective of the researcher (i.e. how to get better data in a more ethical way), Clark grapples with how those researched can experience research fatigue and over-research. Due to this focus on how researchers can deal with over-researched cases and participants, Clark is as concerned with 'research fatigue' as with over-research. As often in discussions of over-research (see more in the following, and this volume), there is a general tendency to elide or at least muddy the distinction between research fatigue and over-research. Here, over-research is a claim made by research subjects suffering from research fatigue, that is, 'we're being over-researched'. However, there are good reasons to hold these two separately.

First, and centrally, over-research can occur without research fatigue. A common example of over-research points out that much social psychology research tends to be with those nearby the researchers. With research often carried out in universities, participants are often reasonably well-educated, well-off, 18–22-year-old undergraduates from a specific class profile, willing to trade an hour of their time in return for some token recompense, often at the same university of the researchers. Yet, just because students and those living near research-intensive universities are over-represented in these studies, it is not always the case that those being studied feel over-whelmed and burdened by their participation—indeed, there is often a steady stream of these ready and willing volunteers. It may even be that overcoming research fatigue can actually help exacerbate the problems of over-research. Clark mentions the lack of change that can result from research projects on hard-pressed groups. Here, overcoming the resistance of research subjects can result in even more research being carried out—overcoming research fatigue can result in over-research.

According to Clark (2008: 955–956), research fatigue is 'a demonstration of reluctance toward continuing engagement with an existing project, or a refusal to engage with any further research'. Clark (2008: 960–961) mentions the possibilities for overcoming research fatigue, including researchers offering the possibility of 'change'—some sort of a material difference in the circumstances of research subjects. Even though this offer of change can be problematic and difficult to fulfil, overcoming research fatigue can even result in more research: over-over-research, if you like. The better researchers and research subjects can get over the fatigue of persistent research encounters, the more research and researchers can mine the same seam of research. Dealing with research fatigue then can be a way to more deeply exploit and extract data from research subjects: akin to fracking participants. In contrast, overcoming over-research involves dealing with the conditions that give rise to the centralisation and focusing in/bunching up of research and hence necessitates a more structural critique of the current state of research.

Over-research is broader than only the fatigue of participants and has implications beyond specific research subjects suffering from research fatigue. These implications include research subjects in general (though this is a sufficient cause for over-research to become a problem), the quality of the research itself, and the narrowing of the expectations of researchers. While there are resources on research fatigue, over-research is a more problematic phenomena, and, due to this more deeply sedimented character, is less likely to be overcome in a voluntaristic manner: that is through better research and training of researchers. Over-research is deeply embedded, and while research fatigue can be pervasive, it is revealed in particularities. Over-research, in contrast, is revealed structurally.

Clark remains focused on analysis at the personal level—albeit shifting from researched to researcher. While personal tips are important, what remains relatively underexplored is how over-research becomes a structural condition of academic research. There appears to be—if not a law—something akin to a general tendency for research fields to pursue similarity, and to tread the same ground, even if done in a slightly different manner. By tracing the distinction between research fatigue and over-research, we can see that over-research is a more structural diagnosis—a whole-system critique of research. If research fatigue is the symptom, over-research is the cause. While clearly related to research fatigue, over-research goes further, is more diagnostic, and a more encompassing term. Focusing on those being researched allows a clear sight of research fatigue but it can miss the wider context of over-research.

Sukarieh and Tannock (2013)

Sukarieh and Tannock expand the focus, from research subjects to a more collective whole community experience of being over-researched. They characterise these over-researched communities as having:

> three key features: communities that are (1) poor, low income, indigenous, minority or otherwise marginalised; (2) communities that have experienced some form of crisis (war, genocide, natural disaster, etc.) and/or have engaged in active resistance to the conditions of their poverty and marginalisation; and (3) communities that are accessible to outside researchers, in particular, by being located in close geographical proximity to research centres and universities.
>
> (Sukarieh and Tannock, 2013: 496)

These need not necessarily be an issue—just because a group is under-resourced or suffering does not mean they are less deserving of research—indeed at times this is a justification. However, over-research occurs not only in places with an over-representation of researchers but also in a context of a modal shift from a beneficial, or even just a light touch, 'do no harm' research approach to one which tips into actively causing or exacerbating the challenges faced by those undergoing research. These problems—taking up time and energy from those researched; the expectations and promises of social change from researchers' promise (even a tantalisingly implicit offer of hope to change their circumstances); the (in)ability of researchers to keep returning to do more research and developing long-term, reciprocal relationships; misgivings over researchers' agendas; and the impact of research within the community—share characteristics with research fatigue and sit alongside what participative action research and other approaches characterise as 'extractive' research.

If this were all to it, the simple solution to the problems of over-research would be for researchers to be instructed to spread out and find cases away from these honeypot sites. For example, see the Urban Studies call to find more 'ordinary cities' (Robinson, 2006) cascading out into a call to focus on more mundane but representative examples, even into the countryside (Kumar and Shaw, 2020). These cases are made less from the perspective of being over-researched but of research more accurately representing a field of study as it is most commonly lived and experienced. But an over-researched place is not so much related to the characteristics of the place—close to a university, or an especially intriguing example, having experienced a high-profile disaster—but to the characteristics of the research process itself. Over-research can also be seen as a structural condition of qualitative (and particularly ethnographic) research as it is currently understood and practised rather than any personal failings or lack of professionalism (or even empathy) on behalf of researchers. Rather, researchers can be doing very good research, be very committed to the causes they are concerned about, and be very nice people to spend time with for those being researched. Researchers can also be scrupulous with the data they collect and have the highest ethical principles in mind when collecting and even when they have collected their data. But we can also note that the same tendencies that lie behind over-research can also be found in research that is not ethnographic in method, not even qualitative. As Sukarieh and Tannock say, 'the solution to community complaints of over-research should not be seen as being solely about moving on to find other, less-researched communities and locations elsewhere'. They also gesture towards the more systemic analysis offered here: 'Traditional, "extractive" research models of university-based and driven research, combined with the continuing proliferation of academic research generally, serve to make the problem of over-research at least a latent concern everywhere' (2013: 507).

As Sukarieh and Tannock (2013) say, unless participative research is connected to a wider project of political and social reform, they are unlikely to cause any material change for those participating. Likewise, there are limits to how much methodological and ethical training of researchers or awareness raising of the problems of over-research can effect change. This is one reason why drawing the distinction between research fatigue and over-research is helpful. Research fatigue can be overcome in particular places but over-research is more structurally embedded in social research.

Sukarieh and Tannock (2013: 507) ultimately argue that it is:

> Only by paying direct and critically reflective attention to the positioning of researchers, research projects, research practices and research institutions within local, regional, national and global structures and processes of power, identity, inequality, interest and control can the problems of over-research and over-researched communities begin to be understood and addressed.

Yet, critical awareness of and action on positionality and power relations brings us beyond the world of research proper and into a wider realm of being political subjects and actors in the world. Such a concern is more properly political and de-centres our identities as researchers, enlacing it within our capacities as political beings. Here, the mode of analysis over-research builds on is not only the more participative form of methodology—particularly their critique of extractive research—but also something more akin to scholar-activism. In particular, the ways that scholar-activism foregrounds the political position and commitments of the researcher.

Neal et al. (2016)

Neal et al. offer two critical additions to this emerging field of literature on over-research. First is a specific focus on place. Second is an argument that over-research is not necessarily a negative aspect, indeed it could be positive.

First, Neal et al. point out that the few expositions that exist on over-research tend to focus on research relationships (i.e. researcher-researched or researcher-researcher) rather than what for them is the key component: 'place-based research relationships'. The over-researched place here is not a passive backdrop that happens to contain over-researched people or processes. Rather, Neal et al. see researchers as active players and participants in the production of a particular place. Thus, research produces *situated knowledges*. It is not that the place contains all these relationships but that it fosters and catalyses these relationships (including relationships with the research) in certain ways. Exploring this role of place in over-research, asks, for example, if it is certain populations or demographics that attract others to certain places or the ways these are enlaced together.

The second key insight from Neal et al. is that over-research may not foster research fatigue, or even result from it, but rather they tantalisingly gesture towards the possibility that over-research may not need to accompany research fatigue at all. This novel addition to the discussion of over-research is among the most comprehensive overviews, perhaps as it is the most recently discussed here. Mainly, this is due to the place anchorage of the article. Yet, there is still the same slippage, if not elision, between research fatigue and over-research. While setting up a clinically separate literature review, and discussing Hackney itself, the lens of research familiarity and research fatigue leaks into something of a proxy for over-research as the focus of the article and analysis. For example, over-research is situated as the fourth aspect of their discussion, placing over-research within a wider frame of research fatigue/research familiarity. While this reflects the way over-research has been approached, pushing further to examine over-research without research fatigue would seem worthwhile.

Neal et al. also hold critically the assumption that participative approaches can overcome over-research, so long as 'research remains extractively orientated' (2016: 4). This short line highlights the need to a more wholesale critique of research practice and orientation rather than a specific researcher's choice of technique or methodology. That much of social research is extractive helps understand the background context in which over-research arises. This is both an extraction of research data from particular empirical cases and sites and also the extraction of intellectual labour from academics towards a publishing-profit complex and impact agenda. There is much of value in this article. Neal et al. centrally argue that the critique of over-research tends to see power as fixed and linear rather than seeing the unequal power relationship as the central injustice of over-research. But if it was possible to develop a research process that empowers research participants, long-term and sustainable forms of research engagement could see over-research not as a negative process but one that provides an insight that more other surface views do not allow.

One reason Neal et al. argue again in moralising over-research—as a 'bad thing'—is the criticism that researchers head to easy reach sites, such as locations that are iconic and media friendly, or in proximity to a university, or even nearby the types of more middle-class neighbourhoods researchers tend to live in, which leads to the saturation of these sites. Such a fusion of justifying case study selection through research-only or research-proper reasons together with extra-professional reasons need not only be a criticism. But, Neal et al. argues creating 'an additional layer of place responsibility and reflexive connectivity' (2016: 7) rather than be attracted to 'iconic. . . [and] on-going sites of research intrigue'. But in preference to swooping in and extracting research before heading back to an affluent neighbourhood, or country, Neal et al. make the case that this holding of the research long after the research encounter allows a 're-knowing' of this experience.

For reasons why certain places become over-researched spaces, Neal et al., through the example of Hackney, add the notion that they are 'symbolic locations'. Taking this term from Gilroy (1987), they argue that these places have a resonance to particular events, practices, or meanings. They are thus more easily amplified—that is used as instructive examples for delivering a wider point or lesson learned. Neal et al. do not explore this aspect of research appeal (that researchers are attracted to these types of case studies/honeypot sites due to the transferability or travelability of the stories they generate). However, I want to argue that this is precisely the reason why we should taking into account the contextual pressures for researchers to make their research relevant and have impact. That is, to have resonance in Gilroy's terms. Likewise, the enjoinment for research to be capable of amplification comes from a background context that cannot be understood solely through individual research choice of a case study or a research subject.

Multiple over-research

While every field has its own examples of over-researched places, drawing these together into something of a general theory of over-research can be a fraught business. As we have seen, over-research is related to and reveals itself in places of research fatigue but there are also some good reasons to hold them separately and hints that they do not have to come together. I have been keen to point to the ways that over-research is a wider, more deeply embedded, and all-encompassing phenomena than particular instances of research fatigue. This also brings in questions of justice and the political position of the researcher. However, the path to an over-researched place, as we can see in the variety of fields in this collection, varies. Some of these reasons are a combination of aspects. The personal vignette at the outset shows researchers starting out potentially unaware that a topic or location has already been studied so comprehensively. The rapidly increasing rate of published research articles has itself increased profoundly in academia. This growth is also uneven, and concentrated in particular outlets, and on particular topics. So, the explosion of interest in particular places can be seen as in part a reflection of the burgeoning outputs generated by academia in general. Being aware of this background environment to gather more research and then publish it is another reason to hold the bigger picture in view. Then, in the second half of this chapter, I want to go big and try and sketch some big picture views that can characterise over-research in the round and what can be done about it.

What can be done?

Building on the cornerstone literature on over-research mentioned earlier, I want to draw out the following arguments: over-research is related to research fatigue in ways that are interlinked but they remain distinct; while descriptions of over-research are increasing, including suggestions of ways to overcome this, explanations tend to *focus on where rather than why or how* research tends towards this centralisation, or bunching up. Where explanations exist, they tend to be rooted in the individual and voluntaristic—and hence tend towards advice to individual researchers, rather than a structural critique of academia/research as a whole. Without appreciating why research tends towards over-researched examples, this advice misses much of its political potential and also remains dealing with symptoms rather than causes of over-research. In this section, I want to outline some of the underlying causes of over-research through weaving together more diagnostic approaches, both postcolonial and political-economic.

Generalist diagnostic approaches

There are a few generalist approaches we can use in diagnosing over-research that outline not only what over-research is or where it tends to take place but also why over-research is a persistent characteristic of research today. These generalist approaches are helpful because they aim to describe research in the round and are thus as much focused on causes as symptoms.

There are many surface generalisations of over-research like researcher's conservatism in pushing new boundaries or researchers sticking to tried, trusted, and dependable examples: 'what works'. Further to these, there are secondary surface reasons—reviewers of both papers and proposals immediately 'get' an example they have heard of before but need to do much more thinking to get their heads around a particularly novel idea or empirical example. New topics also need a lot more explaining which can take away the flow and quick pace of writing. There is also the familiar rhetorical trick in social science of taking an example supposedly well known and giving a new hidden or unseen twist that brings it to life again—much harder to do when introducing something brand new in the first place. With a comparatively under-researched topic, none of this is possible in the same way. This could very easily develop into a 'rut'—or tried and tested good practice, depending on perspective—in a given research track, which deepens and establishes particular examples as the traditional or established way to do research.

Understanding these mechanisms and where the factors that produce over-research come from is important if we are to find a way past this impasse. Digging deeper, we can find more sedimented reasons for over-research. Most of the aforementioned critiques fall into the broad category of bandwagoning. Bandwagoning is basically the idea that someone is more likely to say something, adopt a belief, or carry out a course of action, if it is already well established or legitimised. In methodology textbooks, bandwagoning is normally discussed as a challenge in focus groups: where one or two more dominant figures in the focus group declare something, making it far more likely that the rest of the group will then follow suit or agree. Over-research can be seen as a side effect of a general bandwagoning in any field of research. With bandwagoning well established, it makes sense that it would also play a role in research more generally, with researchers, funders, reviewers, and ethics committees favouring and implicitly accepting the validity of the already assumed or stated axioms. This can be as widespread as a preference for choosing a particular methodology, like a case study approach, to a zooming in on particular place even as it approaches researcher saturation.

There are other more general principles or trends in academia that can be helpful to understand over-research. Some of these are more diagnostic than a general description of bandwagoning: *Matthew effect*, *Price's law*, *Lotka's law*, and *network effects* (see Figure 1.1) (Price, 1976).

The *Matthew effect* describes how being already successful academically is the best predictor for future success: those who are more likely to win the next grant already have one. Those already published are more likely to continue having their articles accepted. Both *Price's law* and *Lotka's law* describe how in any given sub-field or discipline, citation practices tend towards a hierarchy. Each of these is an attempt in some way to describe the gathering or bunching up that happens in various research fields. While these describe citations, these same principles can be seen more widely producing highly authoritative individuals, approaches, experiments, and even places seen as a byword for a certain domain

Bandwagoning: The general tendency for someone to adopt a certain approach or behaviour because others do so. In this case, the tendency to study the same phenomena or place because it is already well established.

Matthew effect: This is the principle that the rich get richer and the poor get poorer (taken from the Bible verse Matthew ch. 25 v. 29). (Merton, 1988).

Price's law: It states that half the literature on a subject will be published by a square root of the total number of authors publishing on that topic. This means that a small minority of authors on a particular topic will be over-represented in the total publications on a topic or sub-field (Price, 1976).

Lotka's law: It is an inverse square law, forming the basis of *Price's law* (Lotka, 1926).

Network effects: The increasing positive utility derived from an increase in connections—in this case, an increase in similar studies or even the same type of study or adopting the same theoretical frame or empirical approach.

Iron law of oligarchy: Robert Michels' theory that any democratic group or organisation tends towards an elite or oligarchy that will come to dominate it.

Figure 1.1 Description of some general principles and patterns in academia that can point towards a general tendency for research to become over-research.

or subfield. There are also *network effects*: when one example or theory is already known, another paper or research project on the same topic more easily fits into and benefits from the increased exposure to that issue.

Each of these tendencies and patterns identify a way of outlining why research practice tends towards a narrow selection of both processes and places. They approach something like a scientific law of over-research: based on repeatable observations, these laws and effects play out in various different fields in a similar fashion. These effects and laws are a different approach to the current literature on over-research because they look to wider patterns of research and so approach something of a general principle or pattern to research. As such they are more structural diagnoses than looking at one of the honeypot sites where this research is actually bunched up. They are important to introduce here although none of these points are interested in research fatigue or individual research experiences from the perspective of the researcher, those researched, or any given over-researched place. This chapter focuses on the selection of where and with whom research is carried out. But this general snapshot of the tendency of research towards hierarchy and a narrowing or bunching up of the questions on *where* research takes places cannot be divorced from outlining the other trends in research that all seem to be of a piece with this general tendency towards a centripetal focus on an ever-narrowing selection in research.

These general tendencies approach something like the *iron law of oligarchy*—for our focus in case studies. The idea is that case study examples will tend to become oligarchic, dominated by a few over-represented cases. Such an approach assumes that these principles are widespread in academic research and that phenomena such as over-research is a symptom rather than a bug of research as currently constituted. Another way to phrase it would be to say that over-research is a structural feature. If this is the case and what we are dealing with in over-research is some form of academic oligarchy, then taking a brief look at postcolonial insights might shed useful light.

Postcolonial critiques

Postcolonialism as an academic study includes analysing the over-representation of particular places and so it makes us see what this mode of analysis has to offer. Emerging from a critique of the enduring ways colonised people and places are controlled and dominated, one strand argues for 'provincialising' former metropoles. Where there was a former imaginary of a natural, neutral, and objective way to view the world, this 'Western gaze' has now been deconstructed by a range of helpful texts, including from Edward Said and Jacques Derrida.

While postcolonialism can mean many different things, two aspects are relevant to our purposes here. (1) Chakrabarty (2008) counters a Eurocentric analysis—the ways in which Europe is taken to be the default base in a range of areas such as cultural analysis, urban planning, and perceptions of beauty—by arguing that any other region in the world (in his case India) can be the cultural equal and should be the locus of theory building as often as 'usual suspects' like Europe. (2) This literature helps us to overcome what could be quite a liberal solution to the problem of over-research. This could argue that while an over-research problem persists, what is needed is not a structural critique but better information to researchers or more finessed techniques in overcoming research fatigue (as mentioned earlier, a side effect of over-research). This can be seen as a naive view that while seeing over-research as problematic, claims that overcoming it can be done merely through moving beyond honeypot sites and diversifying and diluting where theory is built from.

Postcolonialism also helps critique a wide range of processes that can be seen in other fields where a particular, narrow subgroup is taken to stand for a wider whole, at times with deadly consequences. For example, a similar approach can be seen from recent feminist interventions. Criado-Perez (2020) points to the ways the world is designed for men and a particular narrow example of men: as an 'average human'. It is important to note the ways postcolonial, feminist, and ablest critiques outline the catastrophic consequences that can result from focusing one's research attention too narrowly, failing to take into account a wider world. This form of knowledge production sits with the general patterns and tendencies we saw previously and can characterise research today.

Applied to over-research, this perspective would provincialise paradigmatic case studies. It would outline how our theories can be built from places that from a mainstream, dominant perspective can be unthinkable. This will look different in different fields. In urban theory, principles could be built from Begusarai as much as Chicago. Postcolonial perspectives point to the over-representation of certain locations and theories, particularly the central place of Western cases and theories. One key response to this is developing a 'Southern theory' and building new theories from empirical examples beyond traditional sites of knowledge production. Over-research, and tackling over-research, can be a key plank of working towards a more emancipatory research project. It can also be research that better reflects the world as it actually is and not imagined and reproduced from centres of expertise—research that challenges rather than reproduces a centralised, hierarchical society. This perspective is important but going forward, I think we need to also marry Chakrabarty with Bhabha to take a more comprehensive account of over-research.

As mentioned earlier, the real value for me in learning lessons from postcolonialism to help grapple over-research is the avoidance of easy solutions. One such easy solution to over-researched places could be that we merely have to diversify and build theory from unusual places. Bhabha (1994), for instance, cautions against such hope. This 'gesture to the beyond, [i.e. Moving theory-production and empirical data gathering beyond typical places of research/knowledge production] only embody its restless and revisionary energy if they transform the present into an expanded and ex-centric site of experience and empowerment' (Bhabha, 1994: 4). What Bhabha cautions against is the idea of 'beyond'. Finding a new and different place to carry out research because it is different, implicitly entrenches the same processes that produced that difference in the first place. For our purposes, the deliberate choice of unusual or strange places for research is not 'ex-centric' in Bhabha's terms. Rather what is needed is to do away with the 'centric' itself and meet places and phenomena on their own terms. Under this view, the problem of over-research will not be overcome through choosing more novel, original, or curious case study locations. Bhabha's great insight is that it is the ways that certain cases and places come to be seen as unusual, novel, or beyond that is part of the problem. That is, it is the structures that produce over-research, including its accompanying research fatigue that we need to deconstruct, not one specific place or example in itself.

Another way of looking at the issue of over-research is seeing under-research as a shadow side of over-research (see Chapter 7) (Ruszczyk et al., 2021). We can say that awareness of over-research has two phases: first, a revisionist perspective through which more case study locations—diversifying and diluting the over-researched example—can overcome an overly narrow empirical basis for knowledge production. However, resolving many of the concerns for those impacted by social research, most prominently research fatigue, does not get to the more sedimented reasons for over-research. Certain neighbourhoods are no longer swamped with eager young researchers but these eager young researchers—while dispersed—are finding empirical evidence for some sort of universalising theory.

There is a second, more radical rather than revisionist perspective though. This finds places to build theory from and empirical examples, not only because they are what might be expected but also because they are counterintuitive 'instructive examples' and are a site of knowledge production in and of themselves. Neither usual nor unusual suspects, these places 'just are' and can speak to us in their own way. This moves beyond Bhabha's ex-centric critique and towards a more polycentric theory building—provincialising not just over-researched places as they were but dethroning over-research itself. While this analysis has moved in an abstract direction, lessons on how to move forward will need to be particular. There will be many across social science though where I believe these same principles are at work and practical applications of this diagnostic approach can be put to use.

In the following section, I want to take one high profile example of how this polycentric theory building can help move from over-researched places to a more representative selection of places as we build theories. I am aware that by choosing perhaps the most paradigmatic field affected by over-researched places—urban studies—and the most paradigmatic place within it, Chicago, is a little ironic. Heading to Chicago shows how deeply embedded over-research is in social science and also shows potential lessons learned in charting a way out.

Beyond beyond—Chicago as the ultimate over-researched place

Perhaps the most over-researched place is Chicago, particularly for its heritage in urban sociology (Abbott, 1999; Bulmer, 1986; Low and Bowden, 2013). From its production of iconic diagrams, such as Burgess' Concentric Ring Model, to laying some of the cornerstones of urban theory as a whole, the Chicago School of Sociology was one of the most influential and consequential gathering points of sociological, urban, and human ecological research. It is hard to overstate the importance of Chicago to sociology and social science in general. The US's first sociology department in 1892 laid the foundations for what would become urban studies. Key people—such as Robert Park and Ernest Burgess—saw their surrounding city as a laboratory, where universal processes of succession and land use sectors could be observed in detail.

Many of the factors that go into producing an over-researched place can be seen here, at a distance of over 100 years now: easily accessible field sites to the university; iconic social processes unfolding (waves of immigration); and intuitively graspable simplified representations of complex social processes. Yet, there were some other more structural factors too that were outlined in the second section: bandwagoning and a more oligarchic centralisation of what research in this field looked like, who carried it out, and what characteristics it had.

Chicago is a special case. But however special, historic, or interesting, it remains a single case and place. Urban sociology and theory have relied on Chicago as a key research subject in its development trajectory. For example, in 1925, Ernest Burgess developed the Concentric Ring Model, perhaps the most famous diagram in social science. The model was an attempt to explain urban social structures, and like much of Chicago School sociology, used surrounding Chicago as a field site. The Concentric Ring Model outlined concentric rings, centred on a CBD

(Central Business District), enveloped by successive rings of factories, residential and finally commuting zones. The diagram proved an enormous success, simplifying a rapidly expanding and dynamic Chicago. The diagram was a convenient common-sense outline of what Chicago was and then would become. Immigrants arrived and then moved through various 'transition zones'. On top of this basis, various other highly influential models and theories would be built such as the bid-rent curve and it also influenced key concepts like the central place theory. Social science departments worldwide soon came to take these approaches as the default departure point for understanding urban studies and the city.

However, the diagram did not travel as well in practice as it did conceptually. The model did not account well for physical features like mountains, waterfronts, or regions, different cultural settlement patterns, or where there was more desire to live in the city centre, and not pursuing a suburban dream. Cities with different histories, whether old European cathedral cities or planned colonial cities in the Global South, polycentric cities, with various centres, did not quite fit the theory built from Chicago and tweaks and accommodations were made. Indeed, for most cities beyond the US, and even many within it, the model had poor explanatory worth. Later versions of Burgees' model tried to take some of these into account, allowing for the waterfront in Chicago, for example. Subsequent models attempted to improve it, such as the Hoyt model, by taking account of different sectors. But the model made a powerful claim to be the originators, or departure point for these alternative visions.

These additions can be described as 'ex-centric', in Bhabha's language. The problematic replicability or transferability of studies gets to the heart of how over-research can rub up against its limits. Theories and ideas developed from a narrow set of places and then amplified through over-research again highlight larger structural issues. Burgees' concentric model is instructive as it shows that however successful the theories and abstractions we develop from our research are or however close we (think we) are getting towards an isotropic place (an abstract representation a place that will be equally true elsewhere)—and the model has been enormously successful—if the research is limited to a singular, over-researched place, it will inevitably remain partial and particular. Further, it will likely be an echo chamber, an amplification of one particular facet of the world.

This is by no means only a historical argument, and the legacy of Chicago's dominance in social and urban theory still has consequences today, for example, in the ways ghettos have been understood and rhetorically reproduced. Mario Small makes a plea to widen the cases we build urban theory from, appreciating heterogeneity. Discussing the theorising of poorer neighbourhoods, Small traces the ways in which William Julius Wilson's (2012) book *The Truly Disadvantaged* has problematically become the comprehensive theory describing American ghetto development. Small traces how Wilson's persuasive work first developed in Chicago nearly a century after the original Chicago School has then been subsequently fleshed out (and validated) by his students (e.g. Loïc Wacquant, Sundir Venkatesh) and other leading figures (e.g. Eric Klinenberg). Describing US 'so-called' ghettos, Small points out that the dominance of Chicago in understanding ghettos is problematic: the factors and characteristics of a poorer neighbourhood in Chicago just do not function the same way in New York or Atlanta. While the dominant theories of ghettos have been built from Chicago, they travel poorly elsewhere: 'thinking of the problem independent of the theory becomes almost impossible—even when the theory is at odds with the facts' (Small, 2014). Small and Feldman (2012) show in a more detailed manner that the same factors producing the ghettos of Chicago are not always present throughout the rest of the US. As they say, the Chicago studies show that particular is not 'representative' and since it has been predominantly Chicago case studies, and many of the key names in this field who have studied these cases, setting the theoretical agenda on work on poor American neighbourhoods, this presumed archetype has become a stereotype.

While Small's argument applies to the narrow field of disadvantaged neighbourhoods in US cities, Chicago more widely can be seen as a key over-researched place. This reproduced adaptation of theories and ideas developed in Chicago and then applied elsewhere points to the persistence of over-research which, in this case, is more obvious as Chicago again takes centre stage as the over-researched place.

Tracing the history of Burgees' concentric ring model shows one aspect of over-research: that it is a problem for research. The specific problem here was not the impact of Burgees' work on residents of Chicago at the time or the research participants' research fatigue but that the process of abstraction in modelling city from one particular place comes to be the default option. The issue was as much the supposed comprehensiveness of the Concentric Ring Model and its application to understand cities elsewhere. The universality and reification of research have been critiqued elsewhere but they feed into the challenges raising awareness of over-research seeks to address. For anyone trained in urban studies more recently, the Chicago School can seem to be an important stage in the legacy of urban studies but something to be confined to history. Noting the more recent work on ghettos shows how these same patterns of over-researched places coming to dominate a much wider approach still play out, ironically again using Chicago as the default example. As Small argues: 'If all theories must paint a picture of one or another sort, then at least they should ensure that the artist holds a full palette' (Small, 2014).

For Small, this is the full palette. From a postcolonial perspective, the desire is to see beyond the narrow or restricting selection of theory building only from some select, narrow examples.

Conclusion

This chapter has tried to introduce two aspects regularly missing in the discussion of over-research: first, to split over-research from research fatigue. While they regularly accompany each other, both research fatigue and over-research can be found separately. By taking the example of Chicago, we can grapple with the problems of over-research, beyond a consideration of research fatigue, deeply sedimented within the history of social research. Likewise, research fatigue can be overcome to help produce over-research.

Second, this chapter has tried to approach a more general theory of over-research. This more systematic analysis of over-research itself can also be more all-encompassing, including taking into account why over-research is so pervasive in academia today. The downside of this is it can tend to be more general, more abstract, and theoretical. Therefore, the chapter has tried, through the example of Chicago, to show how over-research is a persistent challenge in social research. Though the two examples are a century apart, theories of both urban form and ghettos over-represent and take the lead from Chicago to the expense of applicability and explanatory worth elsewhere.

Importantly, by including perspectives such as postcolonial critiques, the way forward for research is not to softly manage these over-researched places or to try to move beyond seeking out under-researched places but rather to question the processes that produce both over and under-researched places in the first place. Therefore, paying attention to under-researched places simply because they are not the over-researched exemplars is also a problem. It is to move away from, rather than beyond, the usual suspects, the over-researched places, and also from the production of over-researched places.

Finally, I want to dive back down, in particular, to my vignette at the beginning, charting my own subfield of how community is used to pursue environmental values. Research fatigue is clearly an important issue in studying community-based environmentalism. This is one reason why the Transition network produced a highly useful guide for their communities in dealing appropriately with researchers (Bastian et al., 2012). My point in this chapter has been to widen the lens a little though and to separate research fatigue from over-research. In my opinion, research is appreciating all the myriad activities that go under the capacious word community—not just a narrow, select example of easily graspable communities. For almost a decade Transition were seen as a stand in for the many and various ways community is used to respond to environmental challenges, overlooking the capaciousness of community. But understanding how and why such a reified example comes to dominate is as much the task of critical research as finding representative examples to build theory from. Doing so allows an understanding of the general tendencies towards bunching up and hierarchies that can also be found within any given sub-field. The principles and laws outlined earlier—*bandwagoning*, the *Matthew effect*, *Price's law*, *Lotka's law*, and *network effects*, even *the iron law of oligarchy*—are readily relatable within my own field, and, I believe, throughout social science.

References

Abbott, A.D. (1999) *Department & discipline: Chicago sociology at one hundred*. University of Chicago Press, Chicago, IL.

Bastian, M., Brangwyn, B., Giangrande, N., Henfrey, T., Maxey, L. (2012) *Transition research network: New knowledge for resilient futures*. Transition Research Network, Plymouth.

Bhabha, H.K. (1994) *The location of culture*. Routledge, London; New York.

Bulmer, M. (1986) *The Chicago school of sociology: Institutionalization, diversity, and the rise of sociological research*, Nachdr. ed, The heritage of sociology. University of Chicago Press, Chicago.

Chakrabarty, D. (2008) *Provincializing Europe: Postcolonial thought and historical difference, Reissue*, with a new preface by the author. ed, Princeton studies in culture, power, history. Princeton University Press, Princeton, NJ.

Clark, T. (2008) 'We're over-researched here!': Exploring accounts of research fatigue within qualitative research engagements. *Sociology* 42, 953–970. https://doi.org/10.1177/0038038508094573

Criado-Perez, C. (2020) *Invisible women: Exposing data bias in a world designed for men*. Abrams Press, New York.

Gilroy, P. (1987) *There ain't no black in the Union Jack: The cultural politics of race and nation*. Hutchinson, London.

Kumar, A., Shaw, R. (2020) Transforming rural light and dark under planetary urbanisation: Comparing ordinary countrysides in India and the UK. *Transactions of the Institute of British Geographers* 45, 155–167. https://doi.org/10.1111/tran.12342

Lotka, Alfred J. (1926) The frequency distribution of scientific productivity. *Journal of the Washington Academy of Sciences* 16(12), 317–324.

Low, J., Bowden, G.L. (2013) *The Chicago school diaspora: Epistemology and substance*. McGill-Queens University Press, Montreal.

Merton, R. (1988) The Matthew effect in science, II: Cumulative advantage and the symbolism of intellectual property. *Isis* 79(4), 606–623. Retrieved May 22, 2021, from www.jstor.org/stable/234750

Neal, S., Mohan, G., Cochrane, A., Bennett, K. (2016) 'You can't move in Hackney without bumping into an anthropologist': Why certain places attract research attention. *Qualitative Research* 16, 491–507. https://doi.org/10.1177/1468794115596217

Price, D.D.S. (1976) A general theory of bibliometric and other cumulative advantage processes. *Journal of the Association for Information Science and Technology* 27, 292–306. https://doi.org/10.1002/asi.4630270505

Robinson, J. (2006) *Ordinary cities: Between modernity and development, Questioning cities*. Routledge, London; New York.

Ruszczyk, H.A., Nugraha, E., de Villiers, I. (Eds.) (2021) *Overlooked cities: Power, politics and knowledge beyond the urban South*, First Edition. ed, Routledge Studies in Urbanism and the City. Routledge, New York.

Small, M. (2014) No two ghettos are alike. *The Chronicle of Higher Education*.

Small, M.L., Feldman, J. (2012) Ethnographic evidence, heterogeneity, and neighbourhood effects after moving to opportunity, in: van Ham, M., Manley, D., Bailey, N., Simpson, L., Maclennan, D. (Eds.), *Neighbourhood effects research: New perspectives*. Springer Netherlands, Dordrecht, pp. 57–77. https://doi.org/10.1007/978-94-007-2309-2_3

Sukarieh, M., Tannock, S. (2013) On the problem of over-researched communities: The case of the Shatila Palestinian refugee camp in Lebanon. *Sociology* 47, 494–508. https://doi.org/10.1177/0038038512448567

Wilson, W.J. (2012) *The truly disadvantaged: The inner city, the underclass, and public policy*, Second Edition ed. University of Chicago Press, Chicago; London.

2 Overcoming over-research? Reflections from Sydney's 'Petri dish'

Alistair Sisson, Jenna Condie, Pratichi Chatterjee, and Laura Wynne

Introduction

> *It feels like that it's kind of an experiment, which it sort of is; it's like a test case, 'cause it's such a big [project] and yes, there are all sorts of people coming in . . . Kind of chilling, I think, is that some people want to document how many people die because of it. You know it's quite . . . it's quite chilling. And basically the 'petri dish', at the end it'll just get washed out. Also, like the fact that we are under all these microscopes kind of thing, and the government's got a few microscopes in there and so they're kind of going 'OK, they're doing this so we'll throw in this', or, 'we'll throw in some more capacity builders like we'll throw in more staff . . . just see how much it takes to wear them down or just to win'*

These words belong to Karyn, who joined two of us for an interview in 2017, and who we each came to know while researching the redevelopment of the Waterloo estate where she lived. This 18-hectare, 2012-dwelling public housing estate in inner-south Sydney was marked for redevelopment in late 2015. The redevelopment, like so many others around Australia and the world, involved demolishing six modernist towers and dozens of smaller walk-up apartment buildings so that they could be replaced alongside 7000 to 8000 private dwellings. Each of us—Jenna, Laura, Pratichi, and Alistair—were drawn to the place by the injustice of this plan: it was framed by the New South Wales (NSW) Government as a solution to 'concentrated disadvantage' but seemed little more than an effort to capitalise on the enormous value of land in a gentrifying neighbourhood during the peak of the housing boom and erase or dilute a multicultural, low-income community. In 2016, we entered the communities of the Waterloo estate, finding—unsurprisingly—that they, too, were outraged at the redevelopment.

We also quickly found that we were not the only researchers who had been moved to try to understand or intervene in what was unfolding. At that time, the estate was increasingly beset by investigations and investigators of various sorts. There were the four of us, conducting academic research by interviewing, attending meetings and workshops, and so on. But there were also planning consultants, journalists, artists, photographers, filmmakers, architecture students, a theatre group, and a state-sponsored oral historian each of whom made separate but similar demands on the time, energies, and emotions of tenants and community

DOI: 10.4324/9781003099291-3

workers. This is an area that has historically been the subject of a government inquiry, numerous research projects, documentaries, and a previous redevelopment plan. In one meeting, the focus on the area was described as 'vulturish' by a local resident. This is how many research subjects or participants experience and conceive of over-research: not as the presence or performance of too much academic research but as the unsolicited enquiries and unfamiliar faces of all manners of knowledge producers (and extractors), each of whom likely approaches local residents and events with some sense of fascination may tend towards exoticisation and may even seek to identify opportunities for experimentation.

Several scholars have identified the ethical and political problems inherent to conducting research in over-researched places. Perpetual inquisition can cause exhaustion among participants and overstated or false promises about the benefits of research can lead to cynicism and distrust (Sukarieh & Tannock 2013; Clark 2008). Furthermore, in marginalised and disadvantaged places, which are the focus of so many sociological and geographical studies (Crookes 2017), over-research can perform, or at least be experienced as, further scrutinisation and surveillance (Sukarieh & Tannock 2013; Dawson & Sinwell 2012). At its worst, research is bound up in histories of exploitation, criminalisation, colonialism, and abuse (Tuhiwai Smith 2012; Tuck 2018). Understanding over-research from the perspectives of those who are subject to it, as a phenomenon that spans more widely than the field of academia, only heightens the need to consider questions of methodology and ethics.

In what follows, we explain how we grappled with these questions in Waterloo, outlining our responses and drawing lessons from our successes and failures. In the first section, we present some questions which, we argue, researchers should ask themselves when facing such circumstances—questions which can be summarised as: should I research this place? What should I seek to understand about it? And what is my responsibility for the knowledge I might produce? We outline some of the initial questions we asked ourselves vis-à-vis over-research—whether we should do research, what we should research, how we should go about it—and provide some provisional responses. With some qualifications, we highlight the need for new kinds of knowledge production about over-researched places so that they are not consigned to outdated research paradigms or to conditions which have long since passed. We argue that doing research *on* an over-researched place does not necessarily mean only doing research *in* that place, emphasising the importance of academic researchers' abilities to 'jump scales', to situate the local within the regional and the global (Nagar & Geiger 2007). We also advocate for academic collaboration as a way of minimising some of the negative impacts of researching over-researched communities. In the second section, we discuss how we attempted to extend that collaboration into the communities of Waterloo. We outline our scholar-activist approach and contend that the relationships formed through such an approach are vital to establishing an ethical basis for doing research. Yet, as we illustrate, such an approach is a challenging one to adopt. We describe the issues we encountered and examine why we were unable to resolve them. Finally, before a brief conclusion, we attempt to draw some overarching lessons and advice from our experience of doing research in Waterloo, including the need to grapple with questions of procedural versus substantive justice, the importance of open communication, and the need for radical change within our institutions to support engaged, collaborative, and scholar-activist research.

Whether to research, what to research, and how to research?

When we begin to sense that a place is 'over-researched', how should we proceed? What questions should we ask ourselves as researchers? Some simple questions might be: should we pursue research about that place at all? What should we undertake to understand about it? What, if anything, has been left un(der)-examined? And how might we go about investigating these less researched dimensions of over-researched places? How can we balance rigorous empirics, innovations in theory and method, and ethical practice? In one form or another, these were questions that we asked ourselves at various stages of our research in Waterloo. In this section, we offer our somewhat provisional responses.

If a place is over-researched, should we do any new research at all? Having identified a problem, would we then contribute to it? While our chapter (and indeed this collection) implies that researching over-researched places can be justified, these are legitimate questions nonetheless. We should ask of ourselves, as Tuck (2018) implores: is research the intervention that is needed? This is both an ethical question and an epistemological one. Tuhiwai Smith (2012) and others have made it clear that many communities, particularly Indigenous people and other racialised groups, have fraught relationships with research due to ongoing histories of both exploitative and abusive research practices and the weaponisation of research outcomes against them. Research is a dirty word for many such communities, something that has been used to legitimate and reproduce injustices against them. Membership or pre-existing relationships with these communities may alleviate these issues, and in some cases, there are protocols relating to community ownership and determination; nevertheless, as Tuck and Yang (2014: 232) put it, there are still some stories that the academy does not deserve or that it 'has not yet proven itself responsible enough to hear'.

The epistemological question is in one sense about 'who should research' and in another sense about 'how much research is enough?' The latter may depend on research discipline. As qualitative social scientists, we perhaps saw more scope for new research about Waterloo than a demographer or econometrician might see (though perhaps not!). The diversity and breadth of social sciences may lend itself to see more novelty in over-researched places; however, on the other hand, 'harder' sciences hold longitudinal studies in higher regard and may be expected and justified to return more often. Temporal change demands new investigations to understand what has changed and how, for places are always 'in the making' (Neal et al. 2016). We must also strive to understand over-researched places in new ways. It is often the case that the research on a given place has been dominated by a narrow range of disciplines or discourses and that, as a consequence, some aspects or phenomena are ignored or erased. To exclude them from new theory, concepts, methods, and knowledge generally would be analytically flawed and, arguably, unjust. Over-research is not as simple as a threshold quantum; we must also consider the types of knowledge that have been produced and the practices of knowledge production. It should be noted, however, that for local residents of these places who may be asked the same—or very similar—questions by multiple researchers at various times, the experience is frustrating and repetitive. The 'research gap' is not always self-evident to those on the ground.

Waterloo and the neighbouring Redfern are simultaneously over- and under-researched: a large volume of scholarship has been produced but some critically important dimensions have received surprisingly scant research interest. While many Sydney residents would name Redfern if you asked them to identify a gentrified neighbourhood, shockingly little academic research has been published on the process (a notable exception being the work of Wendy Shaw 2000, 2007). Similarly, ask a Sydney resident to draw you a public housing estate and they will likely sketch something that resembles Waterloo; however, the origins of the estate and the reasons for its notoriety have received only minor attention. Furthermore, Redfern is known as the 'crucible' of the movement for Aboriginal self-determination in New South Wales (Foley 2001; Perheentupa 2020). However, only a handful of scholars have engaged with this history in a meaningful way in attempting to understand contemporary urban development and politics (again, see Shaw 2000, 2007). These were some of the lacunae we identified and tried to address.

This leads us to the question of 'how to research'. This, too, is both ethical and epistemological. The knowledge gaps we perceived demanded that we expand the scale of our research to situate Waterloo and Redfern within broader political, economic, and historical structures and processes. While we were interested in and committed to the local struggles in Waterloo, we also approached it as a place to understand neoliberalism and settler colonialism and as a place to critique and contest these structures. This required speaking to politicians and public servants, doing archival research, analysing quantitative data and so on rather than ethnographic work among the community and other local actors on its own. Such 'scale jumping' (Cahill et al. 2007; Nagar & Geiger 2007) is arguably one of academic research's more vital political contributions. Just as importantly, it also reduced the burden of research felt by residents and community workers in Redfern-Waterloo; in other words, it was also a part of our ethical commitment to minimise the extent to which we exacerbated the impacts of researching over-researched places.

The final response to the questions we outlined at the beginning of this section is to encourage collaboration between and among researchers. It is a truism that collaborative research produces better results. Yet, in over-researched places, collaborative research can be a useful way of avoiding some common negative consequences. In a basic sense, collaboration can alleviate our demands upon participants, reducing the likelihood of research fatigue and feelings of exoticisation. It is also an important source of emotional support for researchers in an institutional environment that can be lonely and alienating. The intellectual benefits are important, but it was these ethical considerations, more than concerns about the quality of our research 'outputs', that motivated our collaborative efforts in Waterloo. There are many forms that collaboration can take. It could simply involve preliminary discussions about research aims and objectives. When interviews are involved, it could mean sharing questions so that overlap is minimised and participants are not needlessly repeating themselves or conducting interviews together to reduce the demands upon participants' time. When this is not possible or desirable and participants are being interviewed more than once, previous interviewers might share knowledge and information about what the interviewer should be sensitive to or, alternatively, what they might question further. Each of these were strategies we used in our work in Waterloo. While it is difficult to assess how successful they were in diminishing participants' exhaustion and frustration at the number of inquirers they encountered, we can confidently say it had *some* impact and still produced interviews that had depth and breadth. Establishing a prior relationship with research participants also helped limit the extent to which they felt 'examined' by our presence and our questions. This was a positive side effect of our scholar-activist engagement in Waterloo, which we now discuss in further depth.

From academic collaboration to community collaboration: the productive yet challenging ethics of scholar activism

We each felt compelled, by both our personal research ethics (though not our institutional ethics, as we expand on shortly), to do more than 'just' research in Redfern-Waterloo. The standard operating model—establishing contacts, requesting and conducting interviews and observations, and then disappearing to analyse and report results—seemed neither respectful of residents' circumstances nor an adequate response to their (and our) objections to the redevelopment. From this standpoint, the problem with over-research is not so much the *excess* of research or at least not only that. But the problem lies in how we as researchers go about research. The theories of social change that involve 'documenting damage' and 'raising awareness', so common to the social sciences, lack urgency and, more importantly, are disempowering and ineffective: they constitute a 'colonial theory of change that assumes that it is outsiders, not communities, who hold power to make changes' (Tuck 2018: 159). While we deployed many 'conventional' research methods like interviews, participant observation, and discourse analysis, we also joined a resident action group that had formed for the purpose of contesting the redevelopment process. Members of this group welcomed our involvement, for the most part, given the resources we were able to contribute physically, financially, and symbolically, as well as time, labour, and presence.

Our work with the resident action group occasionally involved contributing to discussions about strategy, tactics, and goals but mostly it was more mundane. As Routledge and Derickson (2015) outline, academic researchers are well placed to expand the capacities of activist groups, and to generate support from other individuals and organisations rather than to tell activists what they should aim for and how they can achieve it (see also Uitermark & Nicholls 2017; Taylor 2014). Among other efforts, we helped manage social media pages, reached out for media attention or wrote pieces ourselves, compiled a handbook about the redevelopment, made submissions to government inquiries and consultations, printed posters and flyers, set up tables and chairs, and so on. Our specialised knowledge and social and cultural capital were important resources but just as important were our bodily abilities, given that many of the group members were elderly. These acts were not intended to be transactional—we hoped to go beyond what Gillan and Pickerill (2012) describe as 'immediate reciprocity' as a way of gaining access—but rather we were acting according to our own political and moral convictions, which aligned with those of the resident action group (at least initially, as we explain in the following). Furthermore, by participating in the mundane logistics of community activism, rather than inhabiting the position of a detached expert or observer, we developed relationships with residents that began to break down the distinction between a scholar and an activist or a researcher and a participant and develop a shared experience (Autonomous Geographers Collective 2010; Featherstone 2003). By developing more meaningful relationships with the residents of Waterloo and becoming 'familiar faces', we reduced, to some extent, the intrusiveness of our presence. Furthermore, in willingly imbricating ourselves into activism, we embrace the research position that 'we do not obtain knowledge by standing outside of the world; we know because "we" are of the world' (Barad 2003: 829). We hoped that by involving ourselves in the place and the lives of its residents, we might both help advance the residents' cause and reduce their sense that they were simply being observed—yet again—by a set of experts descended briefly from their ivory tower.

Being more immersed allowed us to better judge whether and when we should request to interview them and helped transform several of our interviews into a more collaborative dialogue. As Blazek and Askins (2020) describe, it is these relationships between researchers and participants, rather than formal procedures and checklists, in which research ethics are grounded (see also Elwood 2007; Gillan & Pickerill 2012; Bradley 2007). The development of these collaborative relationships is a first step in developing 'situated solidarities' which move us beyond a reflexivity of recognising inequalities of power and privilege so that we might 'accurately' construct and represent the world and towards a reflexivity that enables us to *challenge* these inequalities as well as the very categories of researchers, activists, public housing tenants, etc. (Nagar & Geiger 2007; see also Maxey 1999).

Our scholar-activist approach moves us into the terrain of Rosi Braidotti's 'affirmative ethics', which renders 'the missing people' visible (Braidotti 2019). By 'missing', Braidotti refers to those who are 'empirically missing' because they have been eliminated or 'deselected' (Wynter & McKittrick 2015), such as the sexualised, racialised, and naturalised others. People in an over-researched place can still be 'missing' and struggle for recognition as ways of knowing them are not present in the knowledge systems of neoliberal capitalism, which seek to dehumanise, marginalise, and maintain necropolitical governance. Whether Indigenous, women, queer, poor, non-human animals or technological others, Braidotti (2019) argues that through an ethical praxis that is grounded, accountable, and committed to justice, we can compose a new people and a new community with which to act affirmatively for those whose humanity and knowledge systems have historically been denied and continue to be denied through mainstream 'status quo' knowledge production. Researchers who engage in relational encounters (with people and places) that aim to take down any power differentials can produce affirmation, sustain intensity, and process negativity to forge pathways for a new 'we'. Our affirmative ethics may not have quite extended to 'defining a platform of action on multiple scales in the world' (Braidotti 2019: 161) but our scholar-activism may have helped public housing residents in Waterloo resist being and/or becoming 'missing'. Whether research is really a good platform of action and what it might need to look like conceptually, methodologically, and empirically to be one are questions we come to in the next section.

It is worth noting the instrumental benefits of scholar-activism as well which were at least equal to conventional non-participant ethnographic approaches. We gained access and insights that we might not have otherwise and our ongoing discussions with residents prompted us to rethink our research questions, objectives, and subjects. For instance, we were able to witness first-hand the way in which government actors attempt to destabilise and co-opt community activist groups. Furthermore, several public housing tenants in Waterloo were concerned about their possible transition into social housing managed by a private, not-for-profit provider (public housing in Waterloo is currently managed by the state government). The significance of this transition to tenants prompted two of us to expand the scope of our research and interview people within these organisations. Waterloo residents were also keen to understand the experience of estate redevelopment projects in other parts of Sydney, and other places around the world, and so the relationship between Waterloo and other estates became a larger component of one of our projects, producing insights which we would not have otherwise developed. As Chesters (2012) emphasises, in the context of social movement studies, research participants are knowledge producers in their own right; academic researchers have an ethical responsibility to engage with them as such and there is much to be gained from doing so. Thus, while our primary motivations for our scholar-activist approach were ethical, it also prompted greater depth of engagement with and understanding of our fields of research (though we suspect it might have dissuaded some bureaucrats from agreeing to interviews).

However, there were some immense challenges involved in this approach—issues of appropriation and accountability—and an underlying inequality regarding authority over the terms of engagement and disengagement. In the following, we quote Catherine, another resident of the Waterloo estate, who we became close to during our work in Waterloo.

> *As much as I really liked and appreciated the post-grads and other academics, I felt they could have done more, been more. I do recollect that when a post-grad thoughtfully arranged a teleview with co-author of the very relevant book 'In Defense of Housing' it was introduced as an opportunity for tenants to question the writer. However, with so many post-grads and local personalities lined up with their questions the tenants were side-swiped. It could have been better arranged and tenants respected by being given a space to ask questions without their minders.*

The struggle goes on but the post-grads are not there to witness the latest turn of events.

The situation Catherine describes is one we were anxious to avoid: throughout our involvement in Waterloo, we were concerned about the possibility of speaking over, or for, the tenants who were involved in the action group (see Chatterjee et al. 2019). Undoubtedly, there were times we failed to provide enough time and space for residents' voices. More fundamentally, the communities affected by the redevelopment of the Waterloo estate were not in a position to determine how researchers such as ourselves were to engage with them. No protocols, formal or informal, were in place to govern the level or nature of our involvement; rather, it was up to us to figure it out and decide. Similarly, we were at liberty to walk away without consequence, as we shortly discuss. This is a structural problem in the relationship between research institutions and researched communities, and marginalised communities especially. With some notable exceptions, particularly among First Nation communities, communities like the public housing tenants in Waterloo have not assembled the resources necessary to self-organise and to enforce their terms of engagement and disengagement with research. As such, as much as we might encourage scholar-activism, this inequality between researcher and research subject requires a more fundamental redistribution of capital and power that cannot be resolved by the individuals on either side of this relation.

Our concerns over these unequal power relations often caused us to be too reserved in the presence of conflict or to behave as 'supplicant researchers' (Derickson & Routledge 2015), submitting ourselves to the decisions of the loudest voices in the group. Our fear of appropriating the residents' struggle led us to ignore several problematic group dynamics and increasingly toxic internal politics for too long (again, see Chatterjee et al. 2019). Our presence began to tacitly endorse patronising and undemocratic and at times racist and sexist behaviours of dominant figures in the group. This was an alienating experience for other tenants as it was for us. We found ourselves stuck in a cul-de-sac of a reflexivity focused too intensely on recognising the inequality between ourselves and the tenant members of the group. We felt immobilised, unable to challenge the inequality of our identities, and our silent presence legitimated the discrimination and exclusion that occurred. Consequently, we eventually decided to withdraw, ending our collaboration with the residents' action group. While regrettable, we believed then and now that this was the only remaining ethical option. We knew that we had to step away. The kind of research we were doing was no longer ethically justifiable. We did not, however, know *how* to step away. We quietly withdrew without challenging or confronting the problems we had identified, thinking it was not our place to be part of a solution. As a consequence, and as Catherine politely suggests, we did not remain accountable to the relationships we had formed and the struggle we had committed to. Clearly, we were neither able to break down the distinctions between an insider and an outsider or a community and a researcher nor the unequal power relations between a researcher and the researched. It is one thing to recognise in the abstract that these distinctions are artificial but it is another to overcome them. While we have retained connections to a small number of individual residents, the four of us are now more than likely perceived by residents of Waterloo as yet another set of researchers and onlookers who briefly swept into their neighbourhood for data and then swiftly made an exit as soon as things felt complicated.

Navigating and dismantling the barriers

The challenges we faced in taking a scholar-activist approach to conducting research in an already-over-researched place are similar to those that other researchers have faced (e.g. Creek 2012; Dawson & Sinwell 2012; Sultana 2007). As such, we are inclined to think that they are commonplace and perhaps inevitable. In this penultimate section, we offer some suggestions as to how researchers might provisionally navigate them and, we hope, dismantle them.

Researchers committed to scholar-activist or participatory approaches should think through the tensions between substantive and procedural justice in advance or at the beginning of their participation in pre-existing struggles. Scholar-activist approaches make no pretence at being a passive, unbiased observer; rather, they involve working with participants towards an outcome or goal. The balancing act of seeking particular outcomes through activism versus encouraging and ensuring collective, inclusive, and democratic decision-making can be exceedingly difficult. Uitermark and Nicholls (2017) argue that there are inherent tensions between these two orientations and often trade-offs: with the former, there is a risk that marginalised participants will be yet further excluded; with the latter, there is a risk that the privileged positions we hold as academic researchers are put to waste. In our involvement at Waterloo, we felt concerned that we researchers did not have enough 'skin in the game' to make demands for particular outcomes nor to adopt and implement democratic decision-making structures, and therefore, we did not successfully advocate particularly strongly for either. Whilst this reticence was motivated by the value we placed on procedural justice, our stance was ultimately not productive in shifting the lack of a proper process within the tenant activist group itself. Our fear that we might speak over or for the residents hindered us from helping their cause as we had hoped. If we had been more open and up-front about what we hoped we could contribute to the residents' action group, then we may have felt more confident in pursuing one approach or another or creating a path where the two were in constant conversation with each other.

The importance of open communication generally is another lesson learned. Many residents were unaware we had made the decision to withdraw and while we kept in communication with some, for others we simply disappeared. The decision to leave is not a problem in itself; it is important that researchers care for themselves as well as for others and they should not subject themselves to politics, values, and ethics that conflict with their own (Gillan & Pickerill 2012). As our experience illustrates, the relationship between personal values and those of an organisation or a group is a dynamic one, and therefore one which a researcher must continually evaluate. However, when this relationship becomes untenable, we must endeavour to communicate why: we owe it to the people who have shared our research and activism, and if we do not do it, then there is little chance things will improve. In other words, there is a protocol for leaving, which should be negotiated with the communities in which we are carrying out research (Tuhiwai Smith 2013). Of course, in many cases this can be complicated by group dynamics where particular actors become gatekeepers of communication, making it difficult for researchers to reach out to individuals when things go awry.

There are also many structural, institutional, and logistical barriers to the kind of work that we have outlined and advocated for. As we have discussed, the terms of engagement and disengagement are too often set by researchers themselves in marginalised communities who do not have the resources in place to develop their own objectives and protocols. Furthermore, despite the advantages of our collaboration and scholar-activism in Waterloo, these approaches were far from encouraged by our individual institutions and the wider university sector. While our institutions exhort us to collaborate (!) ad nauseum, they simultaneously impede it in so many ways that the word loses meaning. Solo work is incentivised by performance metrics and demanded by the requirements of research degrees. Ever-increasing workloads make the time-intensive work of collaborating with colleagues and engaging with communities more and more difficult. The hyper-competitiveness of academic labour markets and grant-funding makes many researchers territorial (while others are simply egotistical) with labour directed towards publishing as opposed to supporting our research participants and fellow academics. The kinds of data-sharing which could help avoid participants' sense of being over-researched are often explicitly prohibited by the institutional ethics to which we are required to adhere. Moreover, these institutional ethical procedures largely situate the authority to determine the appropriateness of a project within a small panel of disengaged experts rather than the communities affected by the project. They are overwhelmingly static evaluation processes, carried out prior to commencing a research project. Neither there is little scope to accommodate the change and dynamism of a research site nor is there any protocol to support collaboration between researchers and participants. The pressure to complete research in narrow timeframes, a lack of financial resources for researchers to sustain themselves, the exorbitant tuition fees for international postgraduate students, and the casualisation of academic labour compound such pressures. In these ways and more (e.g. Dawson & Sinwell 2012), the neoliberal university and the labour conditions it imposes, discourage collaborative, engaged, and activist research with communities.

McLean (2018) points out how neoliberal regimes (including the university system) enforce power structures that bury the efforts of activists (including scholar-activists). As such, she calls us to be bolder in our take up of creative and activist research methods influenced by feminist, queer, and arts-based frameworks for a 'planetary urbanism' that traverses communities and other scales. She encourages chaos and risk taking with doses of humility and care. Our ability to do this kind of work depends on our willingness to organise with our colleagues and fight for a university that supports it with job security, fair pay and conditions, manageable workloads, and suitably transformed ways of assessing performance or success (e.g. Connell 2019). This might also involve a radical transformation of institutional ethics procedures. These processes rightly protect the privacy of participants but their prioritisation of risk minimisation *for the university itself* (Dekeyser & Garrett 2018) can lead to restrictions on researchers that prohibit working with and among communities and activist groups. This can place the ethics of scholar-activists at odds with institutional ethics: institutional ethics requires that we identify the 'potential benefits' of our research according to the colonial theory of change—raising awareness to bring about change from above (Tuck 2018; see also Tuck & Yang 2014)—while scholar-activist ethics follow theories of change that centre communities and solidarity (Cahill 2007; Bradley 2007; Martin 2007). In addition, institutional ethics procedures also do not offer much guidance or support when things go badly (Elwood 2007). Institutional ethics should not be a barrier to participatory and scholar-activist research but rather a resource for non-extractive research, assisting researchers to develop such practices and supporting them when things go awry.

Conclusion

The questions prompted by over-researched places are not limited to those of 'whether' we should do research. Important as these questions are (and assuming an answer in the affirmative), we must also ask questions relating to 'what' we research and 'how' we do it. While more research is not always justifiable, we should be wary of excluding over-researched places from new questions, theories, concepts, methodologies, and epistemologies, for ethical as much as empirical reasons. At the same time, we should recognise that over-researched communities are often frustrated, cynical, or fatigued from researchers' demands and we need to strategize methods that avoid or minimise these effects. Our preceding discussions suggest two changes to academic research that can support the wellbeing and agency of communities, especially in contexts of over-research. First, collaboration among academic researchers can help reduce the demands on participants and enable researchers to be more attuned to how individuals experience the research process. Collaboration with communities themselves can move research further from extractivism by contributing to their objectives and goals. Where communities are involved in political struggle, scholar-activism provides a framework for such collaboration. We took this approach in working closely with the residents' action group in Waterloo. We began to develop the 'situated solidarities' (Nagar & Geiger 2007) that oriented our research towards *challenging* the processes in which research participants found themselves. We built a more compassionate and nuanced understandings of the research process itself and of the challenges that arise when engaging in both activism and research. Second, we call for a radical transformation within institutional ethics and the nature of university work, in general, such that researchers and their research can support transformative social change. The relationships between participants and researchers are fraught and limited: our own experiences attest to the challenges in navigating these relationships, including the structural inequality where marginalised communities do not have the protocols for engagement and disengagement with researchers. University ethics procedures are not set up to manage this context. They situate the authority to define, evaluate, and prescribe ethics with detached academic elites. A restructuring of these institutional processes in line with the scholar-activist values can help reframe ethics from a procedural checklist to a dynamic 'dialogue with research participants and thoughtful mutual engagement with [ethical] dilemmas when they occur' (Elwood 2007: 337). The collaborative, scholar-activist approach comes into conflict with institutional ethics protocols which situate authority within detached academic elites. We therefore conclude by reiterating the need for radical transformations within institutional ethics and the nature of university work in general, so that researchers and their research can support transformative social change.

References

Autonomous Geographers Collective. (2010). Beyond Scholar Activism: Making Strategic Interventions Inside and Outside the Neoliberal University. *ACME: An International E-Journal for Critical Geographies*, *9*(2), 245–275.

Barad, K. (2003). Posthumanist Performativity: Toward an Understanding of How Matter Comes to Matter. *Signs: Journal of Women in Culture and Society*, *28*(3), 801–831.

Blazek, M., & Askins, K. (2020). For a Relationship Perspective on Geographical Ethics. *Area*, *52*(3), 464–471.

Bradley, M. (2007). Silenced for Their Own Protection: How the IRB Marginalizes those it Feigns to Protect. *ACME: An International Journal for Critical Geographies*, *6*(3), 339–349.

Braidotti, R. (2019). *Posthuman Knowledge*. Cambridge: Polity Press.

Cahill, C. (2007). Repositioning Ethical Commitments: Participatory Action Research as a Relational Praxis of Social Change. *ACME: An International Journal for Critical Geographies*, *6*(3), 360–373.

Cahill, C., Sultana, F., & Pain, R. (2007). Participatory Ethics: Politics, Practices, Institutions. *ACME: An International E-Journal for Critical Geographies*, *6*(3), 304–318.

Chatterjee, P., Condie, J., Sisson, A., & Wynne, L. (2019). Imploding Activism: Challenges of Housing Scholar-activism in the Neoliberal City & University. *Radical Housing Journal*, *1*(1), 189–204.

Chesters, G. (2012). Social Movements and the Ethics of Knowledge Production. *Social Movement Studies*, *11*(2), 145–160.

Clark, T. (2008). 'We're Over-Researched Here!': Exploring Accounts of Research Fatigue within Qualitative Research Engagements. *Sociology*, *42*(5), 953–970.

Connell, R. (2019). *The Good University: What Universities Actually Do and Why It's Time for Radical Change*. Zed Books Ltd.

Creek, S. J. (2012). A Personal Reflection on Negotiating Fear, Compassion and Self-Care in Research. *Social Movement Studies*, *11*(2), 273–277.

Crookes, L. (2017). The "Not so Good", the "Bad" and the "Ugly": Scripting the "Badlands" of Housing Market Renewal. In P. Kirkness & A. Tijé-Dra (Eds.), *Negative Neighbourhood Reputation and Place Attachment: The Production and Contestation of Territorial Stigma* (pp. 81–101). London & New York: Routledge, Taylor & Francis Group.

Dawson, M. C., & Sinwell, L. (2012). Ethical and Political Challenges of Participatory Action Research in the Academy: Reflections on Social Movements and Knowledge Production in South Africa. *Social Movement Studies*, *11*(2), 177–191.

Dekeyser, T., & Garrett, B. L. (2018). Ethics ≠ law. *Area*, *50*(3), 410–417.

Derickson, K. D., & Routledge, P. (2015). Resourcing Scholar-Activism: Collaboration, Transformation, and the Production of Knowledge. *The Professional Geographer*, *67*(1), 1–7.

Elwood, S. (2007). Negotiating Participatory Ethics in the Midst of Institutional Ethics. *ACME: An International Journal for Critical Geographies*, *6*(3), 329–338.

Featherstone, D. (2003). Spatialities of transnational resistance to globalization: The maps of grievance of the Inter-Continental Caravan, *Transactions of the Institute of British Geographers*, *28*(4), 404–421.

Foley, G. (2001). Black Power in Redfern 1968–1972. *The Koori History Website*. www.kooriweb.org/foley/essays/pdf_essays/black%20power%20in%20redfern%201968.pdf

Gillan, K., & Pickerill, J. (2012). The Difficult and Hopeful Ethics of Research on, and with, Social Movements. *Social Movement Studies*, *11*(2), 133–143.

Martin, D. G. (2007). Bureaucratizing Ethics: Institutional Review Boards and Participatory Research. *ACME: An International Journal for Critical Geographies*, *6*(3), 319–328.

Maxey, I. (1999). Beyond Boundaries? Activism, Academia, Reflexivity and Research. *Area*, *31*(3), 199–208.

McLean, H. (2018). In Praise of Chaotic Research Pathways: A Feminist Response to Planetary Urbanization. *Environment and Planning D: Society and Space*, *36*(3), 547–555.

Nagar, R., & Geiger, S. (2007). Reflexivity and Positionality in Feminist Fieldwork Revisited. In A. Tickell, E. Sheppard, J. Peck, & T. Barnes (Eds.), *Politics and Practice in Economic Geography* (pp. 267–278). London & Los Angeles: SAGE.

Neal, S., Mohan, G., Cochrane, A., & Bennett, K. (2016). 'You Can't Move in Hackney without Bumping into an Anthropologist': Why Certain Places Attract Research Attention. *Qualitative Research*, *16*(5), 491–507.

Perheentupa, J. (2020). *Redfern: Aboriginal activism in the 1970s*. Canberra: Aboriginal Studies Press.

Routledge, P., & Derickson, K. D. (2015). Situated Solidarities and the Practice of Scholar-activism. *Environment and Planning D: Society and Space*, *33*(3), 391–407.

Shaw, W. S. (2000). Ways of Whiteness: Harlemising Sydney's Aboriginal Redfern. *Australian Geographical Studies*, *38*(3), 291–305.

Shaw, W. S. (2007). *Cities of Whiteness*. New York: John Wiley & Sons.

Sukarieh, M., & Tannock, S. (2013). On the Problem of Over-researched Communities: The Case of the Shatila Palestinian Refugee Camp in Lebanon. *Sociology*, *47*(3), 494–508.

Sultana, F. (2007). Reflexivity, Positionality and Participatory Ethics: Negotiating Fieldwork Dilemmas in International Research. *ACME: An International Journal for Critical Geographies*, *6*(3), 374–385.

Taylor, M. (2014). 'Being Useful' after the Ivory Tower: Combining Research and Activism with the Brixton Pound. *Area*, *46*(3), 305–312.

Tuck, E. (2018). Biting the University that Feeds Us. In M. Spooner & J. McNinch (Eds.), *Dissident Knowledge in Higher Education* (pp. 149–167). Regina: University of Regina Press.

Tuck, E., & Yang, K. W. (2014). R-Words: Refusing Research. In D. Paris & M. T. Winn (Eds.), *Humanizing Research: Decolonizing Qualitative Inquiry with Youth and Communities*. London: SAGE Publications, Inc.

Tuhiwai Smith, L. (2012). *Decolonizing Methodologies: Research and Indigenous Peoples*. London: Zed Books.

Tuhiwai Smith, L. (2013). *INQ13 | Linda Tuhiwai Smith and Eve Tuck—"Decolonizing Methodologies"*. 29 April, viewed 28 September 2020. www.youtube.com/watch?v=rIZXQC27tvg.

Uitermark, J., & Nicholls, W. (2017). Planning for social justice: Strategies, dilemmas, tradeoffs. *Planning Theory*, *16*(1), 32–50.

Wynter, S., & McKittrick, K. (2015). Unparalleled Catastrophe for Our Species?: Or, to Give Humanness a Different Future: Conversations. In *Sylvia Wynter: On Being Human as Praxis*. Durham: Duke University Press.

3 Epistemological, decolonial, and critical reflections in constructing research in former Yugoslavia[1]

Cyril Blondel

Introduction

> In order to expose and undercut this reinscription of otherness, research on East-Central Europe should engage with postcolonial theory.
>
> (Kuus 2004, p. 472)

While it is difficult to establish that the Balkans are an over-researched area, they are often approached in the same (limited) way: reduced to its post-imperial, post-socialist, and post-ethnic conflict past, stuck in history, and orientalised. As if the West was not concerned by post-imperialism, postcolonialism, post-socialism, and by its (still ongoing) responsibilities in World conflicts. This self-imagination of clean modernity impacts Western researchers when approaching the Balkans. It is not problematic *per se* as no one can entirely objectify themselves, and get out of themselves. Nevertheless, I argue in this chapter that approaching a space as fantasised as the Balkans requires one to carefully reflect upon his/her own position towards these over-used categories to better differentiate what is their own analysis about the processes they attempt to research, from what is their own participation in those processes.

In 2004, Merje Kuus, an Estonian geographer based in Canada, invited researchers working on the process of the enlargement of the European Union (EU) (then towards Central and Eastern Europe) to integrate the contributions of postcolonial theory into their reflections. This chapter aims to highlight the relevance and timeliness of such assertion for someone who started a thesis on the pre-accession policy at the Serbia/Croatia border in 2008 (Blondel 2016). It is also the story of the adaptation of such advice to my specific work, that is, as much to the supposed specificities of the regions (and populations) researched as to the place of enunciation of this research (a social sciences research centre in a *provincial* French university) and its enunciator (I, a French white man).

The first adaptation is of a spatial nature: almost all the countries that have been concerned by enlargement since 2008 are located in Southeastern Europe. They belong to a geographical group that the EU names *Western Balkans*. These are the new territories of European expansion (Blondel 2013). The second adaptation is

DOI: 10.4324/9781003099291-4

of a theoretical nature: decolonial thought has emerged to supplement the contributions of postcolonial theory. Among its main contributions, we note the criticism of the hegemonic attitude of the West, including its thinkers, towards the non-West, and particularly its periphery and semi-periphery (Tlostanova 2015).

This point of departure constitutes a statement of position, that of considering that research work 'is subject to the relations of knowledge and power that have a history[2]' (Fassin 2008a, p. 318). This also reflects an intention: 'rather than evading them, we must try to understand them, and, for example, to question the very conditions that make an ethnography conducted far from home possible today' (*ibid.*). Thus, the arrangements of my entry into the world of the Serbia/Croatia border are at the same time part of these socio-spatial relations and the developing agent which allows me to try to apprehend and understand them (Gaillard 2013). If I paraphrase Didier Fassin (2008b), critical analysis of the ethnographic situation—as the historical and geopolitical stage where the encounter between the researcher and his interlocutors—and the ethnographic relationship—as an unequal relationship, in both directions, which is formed between the investigator and the respondent—then constitutes what makes knowledge in social sciences possible (Fassin 2008b). This is precisely the purpose of this chapter: to return to the epistemological and political stakes posed by my research. It is a question here more particularly of updating the issues raised and the difficulties encountered in delimiting the subject: temporally, historically, and territorially.

The question posed in my thesis work, that of the developments and persistence of the socio-spatial relations on the border between the two nation states, was strongly marked by the political context of pre-accession to the EU. The latter enjoins the two countries and their respective populations to reconcile by cooperating (and cooperate by reconciling). This framework influences the way in which the issue of enlargement is historically, geographically, and normatively posed by (Western-European) 'cultural producers' in general (researchers, journalists, and politicians) (Wacquant 2007; Wacquant et al. 2014) and by me in particular. The post-Yugoslav space is indeed treated as a separate sub-field, governed by specific themes. This trend is perpetuated by institutional funding and research strategies in which academics are directed to (and choose to) enrol (Jansen 2015).

When we focus on theses concerning the region submitted to French universities since 2005, we find subject-area registration mainly in the field of law and political science. We note a concentration of subjects on two sometimes related fields: (1) nationalisms, conflicts, and their consequences and international justice; (2) European enlargement, its mechanisms, challenges, and effects (Blondel et al. 2015, p. 8). This reveals the dual perspective (problem/solution) and dominant view of the region. On the one hand, it is approached according to the potential danger that it would continue to represent for peace in Europe. On the other hand, we measure if and how it manages to normalise itself to become Europeanised (*ibid.*). One wonders to what extent this reading really contributes to the understanding of the socio-spatial and political phenomena that are currently at work and to what extent it does not contribute rather, or at least at the same time, to perpetuating them. Might not this portrait be reductive and does it not say at least as much about the phenomenon observed (post-Yugoslav changes) as about those who produce it (and especially about us, the researchers, and more broadly about the academic situation)? What then are the main epistemological and political questions posed by the ethnographic situation—that is, by the relationship between the researcher and his or her fieldwork?

To answer this question, I will return in the three first sections of this chapter to three dominant ways of approaching this kind and this type of questioning in social science research: (1) through nationalisms, (2) through post-socialism, and (3) through the post-Yugoslav paradigm. This is not to question here the fact that researchers have an angle of approach, a bias. I share the point of view of Žižek (1994): any position is ideological. The idea here is rather to discuss the limits of the dominant approaches in the particular case of my thesis by trying to identify as much the postulates on which they rest as the blind spots that they produce. In this way, my goal is to contribute to a broader and more general reflection on the conditions for producing knowledge and the validity/relativity of knowledge. I will then address, in a final part (4), a less-conventional approach, the decolonial option, what it brought me in the context of the thesis, and also its significant contribution to the reflections on research into research.

These mental gymnastics allowed me to occasionally overcome—but regularly only report on—the limits of my research but also to recognise the unsurpassability of certain aspects of my work connected to the situation in which I stated and conducted it. Reflecting on epistemological reflexivity contributes to clarifying the scientific (in)validity of research subject positions and critically situates researcher's arguments and their position and positionality. Doing so, this chapter is firmly set up as a reflective and theoretical piece of research that aims at calling for alternatives to over-dominant *modern* Western epistemology.

Escaping methodological nationalisms

The post-Yugoslav space is today most often tackled through its nationalisms, that is to say, by postulating these as characteristic social and spatial facts. However, this angle of approach can be a cognitive trap: faced with the challenge of observing nationalism in the field, the researcher often comes to nationalise his or her view, which is commonly called methodological nationalism. It was against the danger of the nationalist reduction of my view of the Serbia/Croatia border situation that I first conceived as my approach in the field.

As Sperenta Dumitru (2014, p. 18) reminds us, the critique of methodological nationalism is indeed an epistemological question, in the sense that it is 'neither to defend nor to represent globalisation, the collapse of the nation-state, or borders' but to raise 'a question of methodology of social science research'. According to her, we find mainly three forms. All of them are symptomatic of the tendency to over-research nationalisms in former Yugoslavia.

The first, 'stato-centrist' nationalism, leads to granting an unjustified pre-eminence to the nation-state, whether in social or political analysis (Beck 2003, p. 62), as if law and social ideals were only defined by the state and existed only through it (Dumitru 2014). Without denying the influence of national variables, it is appropriate, instead, to make them into one variable among others—along with Europe and local [variables] in particular—in the analysis of the creation of the border (Pasquier 2012). 'Among others' means that it is equally important not to fall into a locked-in approach on other scales. Although less frequent, the risk of excessive European or localistic tropism would also be problematic. The border reconfigurations (in the case of my thesis, Serbo-Croat) do not operate exclusively on these two scales either; besides, 'the use of internal/external and national/international polarities served to hide the interaction between processes taking place on different scales' (Agnew 2014). Thus, avoiding stato-centrism requires thinking on different scales, but above all, grasping what is happening between scales.

The second form is that of methodological nationalism, called 'groupist' in reference to Rogers Brubaker—it consists of understanding (and reducing) society to that of a nation-state (Dumitru 2014). Specifically, distinct presupposed groups, clearly differentiated, internally homogeneous, and delineated externally, are considered as basic constituents of social life, chief protagonists of social conflict, and fundamental units of social analysis (Brubaker 2002). This tendency is frequent in the study of national, racial, and ethnic conflicts; especially, when we talk about Serbs and Croats in former Yugoslavia by reifying them as substantial entities to which interests and *agencies* (in the sense of Pierre Bourdieu 1991) can be attributed (*ibid.*). In the case of my thesis work, becoming detached from 'groupism' has required the consideration of ethnic belonging as non-homogeneous and not exclusively Serbian or Croatian and social affiliations as not exclusively ethnic and national. Once again, it is a question, without abandoning it, of not giving too much importance a priori to the ethnic analytical model but making it one among others according to what individuals mention.

The third form of methodological nationalism that was identified is termed 'territorialist'. It comes down to 'understanding space as naturally divided into national territories' (Dumitru 2014, p. 22). In the analysis, such a perception leads to formulating concepts, asking questions, constructing hypotheses, collecting and interpreting evidence, and drawing conclusions in a spatial framework that is completely territorialised (Scholte 2000). This is what many authors have called the 'territorial trap' (Agnew 2014; Hadjimichalis & Hudson 2007). In the same way as 'groupism' at the social level, this tendency leads to the reification of 'state territories into given or fixed units of sovereign space', which is equivalent to 'dehistoricising and decontextualising the processes of the creation and disintegration of States' (Agnew 2014, p. 30). In the case of my research, trying to avoid the territorial trap led me to think of the Serbia/Croatia border in other ways (e.g. spatial and reticular). This led to an attempt to avoid the confinement of fieldwork, as much as analysis, in the cross-border territory, as defined and prescribed by the European Programme for Cross-Border Cooperation (2007–2013). It was also a matter of monitoring the observation of the border and the projects organised around it outside of the territorial frameworks and a pre-imposed lockstep.

Dumitru rightly points out that almost no research avoids the three forms of methodological nationalism, especially as they are articulated. Indeed, the idea of the state refers to a certain form of verticality in social and territorial interlocking. James Ferguson and Akhil Gupta perfectly summarise the logic:

> Verticality refers to the central and pervasive idea of the state as an institution somehow 'above' civil society, community and family. . . . The second image is that of encompassment: Here the state (conceptually fused with the nation) is located within an ever-widening series of circles that begins with family and local community and ends with the system of nation-states. This is profoundly consequential understanding of scale, one in which the locality encompassed by the region, the region by the nation-state and the nation-state by the international community. These two metaphors work together to produce a taken-for-granted spatial and scalar image of a state that both sits above and contains its localities, regions and communities.
>
> (Ferguson & Gupta 2002, p. 982)

What Ferguson and Gupta describe is also what is happening in the former Yugoslavia: over-searching nationalism(s) is tantamount to overestimating the political, social, and territorial importance of the state.

In my thesis work, to avoid methodological nationalisms, I tried, as much as possible, to approach the border phenomenon not as an interstate (between Serbian and Croatian states), inter-national (between Serbian and Croatian 'peoples'), or even inter-territorial process (between a nesting of Serb territories and a nesting of Croatian territories) but as a socio-spatial configuration (in its daily routine and by the injunction to cooperate) that challenges precisely all these categories. The challenge was to take into account the relationship as much as the disconnect between the space and scale (Neveu 2007), that is, the transnational character of both the state and the local (Ferguson & Gupta 2002) without falling into the 'methodological fluidism':

> While it is important to push aside the blinders of methodological nationalism, it is just as important to remember the continued potency of nationalism. Framing the world as a global marketplace cannot begin to explain why under specific circumstances not only political entrepreneurs, but also the poor and disempowered . . . continue to frame their demands for social justice and equality within a nationalist rhetoric.
>
> (Wimmer & Glick Schiller 2003, p. 600)

In my case, more than the (post-Yugoslav) states themselves, it is the ideologies attached to them that were at the centre of my attention. Does the ideology of reconciliation replace nationalist ideologies, strengthen them, or accommodate them? The objective was to try to better apprehend the socio-spatial relations related to other changes, other permanences, and other anchorages than those of the State and the nation. Again, my point here is not to assert that the former-Yugoslavia region is over-researched *per se* but to claim that it is over-researched in certain ways. Ildiko Erdei noted a certain fatigue in the eyes of researchers on this space, especially in the mobilisation of the dichotomy nationalism/anti-nationalism as a unique 'analytical vocabulary and explanatory paradigm' (2009, p. 82). In my thesis work, it was a matter of following her invitation to further mobilise other explanatory paradigms to approach and better understand 'new' post-conflict, post-Yugoslav, and post-socialist societies.

If leaving the explanatory paradigm of nationalism/anti-nationalism is a first step, then this work is far from sufficient and one ideology replaces another. And it would be naïve and vain to think that one could develop a language that is theoretically neutral and non-biased: 'While we are still striving for an adequate terminology not coloured by methodological nationalism, we can already predict that emerging concepts will necessarily again limit and shape our perspective, again force us to overlook some developments and emphasize others' (Wimmer & Glick Schiller 2003, p. 600). Each observation depends on the positionality of the researcher. Each of his or her analyses depends on the conceptual focus which limits the scope of the empirical research and the interpretations. The challenge is then to position his or her research theoretically and epistemologically to find the balance between intelligibility and consistency: 'the task is to determine what reductions of complexity will make best sense of the contemporary world and which ones are leaving out too many tones and voices, transforming them into what model builders call *noise*' (*ibid.*).

So, beyond methodological nationalisms, what are the other dominant 'conceptual structures' put in place to address the post-Yugoslav space? What are their contributions and their limits? And how have I positioned myself epistemologically with regard to them?

Going beyond the reading of the post-socialist transition

The second dominant reading of post-Yugoslav space, and more broadly of European states formerly belonging to the 'communist bloc', is offered by the supposedly explanatory paradigm (but just as normative as the methodological nationalisms) of post-socialism. It led to describing former Yugoslavia as facing the challenge of 'democratic transition' (Chiclet 1997), in the process of *Europeanisation* (Štiks 2009), in the process of an unfinished stabilisation process (Dhorliac 2014). These few formulations reveal how the region has often been portrayed in French research (and not only therein) in recent years. However, as political scientist Jean Leca points out,

> it is difficult to distinguish what in the field of *transitology* relayed by *consolidatology* is based on the empirical analysis of a process about which the scientist theorises, and which concerns participation in a process in which the scientist theorises and the citizen acts.
>
> (Leca 2000, p. 108)

This is the first limit of these approaches; the lack of reflexivity in the text does not allow us to distinguish clearly what the explanation is of and what comes under the prescriptions to 'democratise', 'to Europeanise', and to 'stabilise'.

Such approaches underlie, by the language used, a conceptual structuring in 'post-', post-conflicts and post-nationalism but also post-socialism, post-communism, and even post-Yugoslav (I will come back to the last one in the next section). The pre-accession process for the EU, as it is conceived today by the Western Balkans, largely takes up the precepts conceived in the context of the accession of the countries of Eastern Europe, strongly influenced by a post-socialist reading. This reading raises a question: 'Postsocialism gets lost because it is largely presumed to be a process of democratization or Europeanization and thus uncritically positioned vis-à-vis the first World' (Suchland 2011, p. 839). Research on the democratic transition then becomes an implicit field of comparison of which the West apparently constitutes the standard, implicit or explicit: 'the models of transformation observed in the consolidated hyperreal democracies of Western Europe are treated as the only valid model for democracy. Actors and structures found in other societies are signified as deficits of or obstacles to democratization' (Boatcă & Costa 2010, p. 22).

Indeed, the 'post-' approach is based on a generally binary reading that postulates in the first place a territorial confinement between the two so-called homogeneous and opposite blocs (socialist bloc versus capitalist bloc, democratic bloc versus nationalist bloc, etc.). Madina Tlostanova (2012, p. 131) questions, for example, this supposed homogeneity in post-communist categorisation: 'Postcommunism itself is a highly questionable umbrella term lumping together societies which share an experience of communist political regimes but have different local histories and distinct understandings of their situation, aims, roles and prospects in the global world' as much as Jennifer Suchland (2011, p. 844): 'we cannot safely say . . . that the post-communist space is or was a homogeneous place'. For this reason, an alternative take requires to start empirical research from the ground and from grounded histories by postulating that each context is a different form of appropriation of socialism—which seems all the more visible in the case of the Yugoslav, the so-called *third way*.

Beyond that, this reading also postulates a temporal break on which a narrative of modernity is based—everything was bad before in your traditional model (*socialist*, *Yugoslav*, *Balkan*, and *nationalist*), everything will be better in the future if you follow our progressive model (*European*, *liberal*, and *democratic*): 'transition is perceived as not only a necessary, but also a well-defined, clearly directed process at whose end the former socialist societies should fully implement ready-made models coming from the West' (Petrović 2014, pp. 10–11).

Temporal categorisation then serves the differentiation according to a scale of progress: 'The *catching up* timeline can be seen as temporal othering, based on a linear conception of temporality that generates a periodisation of chronological sequences and functions as a taxonomy of progress and backwardness' (Koobak & Marling 2014, p. 338). This differentiation gives rise to a balance of power between the situations observed: 'difference is understood as points on a vertical scale of inferiority/superiority, presence/lack or advancement/ backwardness, rather than on a horizontal field of plurality in which no point has definitional advantage over the others' (Sarkar 2004, p. 328). The standard narrative of a West European modernity represents the 'colonisation of space by time' (Tlostanova 2010, p. 21), 'the obliteration of space by time' (Koobak & Marling 2014, p. 338), or the 'discursive victory of time over space' (Massey 1999, p. 31) of which Doreen Massey offers a summary portrait: 'That is to say that differences that are truly spatial are interpreted as being differences in temporal development—differences in the stage of progress reached. Spatial differences are reconvened as temporal sequence' (*ibid.*).

'Post-socialism' is not only a geographical and temporal label but also a Western-centred analytic category (Koobak & Marling 2014, p. 334). Its use can lead the researcher to participate in the reproduction of the balance of power on which this modernist reading of the world, and in particular of the Balkans, is based. Critics of this reading are numerous. Summoning certain contributions of the postcolonial approach, Todorova's decolonial works (1997, p. 60) and Immanuel Wallerstein's (2014) theory of World-Systems, Manuela Boatcă, for example, underlines the symbolic violence in the body–(semi)periphery relationship between Western Europe and the Balkans:

> Geographically European (by 20th century standards, at any rate), yet culturally alien by definition, the Balkans, as the Orient, have conveniently absorbed massive political, ideological and cultural tensions inherent to the regions outside the Balkans, thus exempting the West from charges of racism, colonialism, Eurocentrism and Christian intolerance while serving 'as a repository of negative characteristics against which a positive and self-congratulatory image of Europe and the West has been constructed.
>
> (Boatcă 2006, p. 327)

The Balkans embody in the European imagination, the geographical, temporal, and symbolic gap between the West and the East, that is to say both the convenient margin that is invoked as a negative reference and the shield that protects from much worse.

If we apply the paradigm of transition to the field of research, then the Serbo-Croatian border represents as a place of evil, partly fantasised, and a space (among others) to integrate into modernity, the first consideration serving to justify the second. The process of enlargement of the EU to the Western Balkans is the continuation of the work to absorb the East European:

> this new *civilizing mission* meant being once again defined as *catching up* with the West and embarking on a supposed transition from the Second to the First World, whose conditions—in the form of EU regulations. . .—are being dictated by the latter.
>
> (*ibid.*, p. 340)

The discourse of modernity is carried and reproduced by the EU, which defines the entry standards to its body but also by the candidate countries that aspire to integrate the centre:

> Politically and epistemologically, what is at stake for those ex-communist countries having long made the bone of contention of Europe's powerful

> empires is the possibility of a renewed shift of axis—away from the semi-peripheral identity of an Eastern bloc country and toward a yet-to-be-defined position within the orbit of the Euro-American core.
>
> (*ibid.*)

This pirouette from the East to the West is particularly visible in the discourse of the elites in, for example, Slovenia, Croatia, or Estonia: the constant rejection of their Orientality and the accent placed on the opposite of their will, and even their right, to Westernisation seen as a 'return to Europe' (*ibid.*; Blondel 2021).

Under the guise of analysing—and enjoining—Europeanisation, many researchers reproduce this discourse and this posture

> much of . . . research, both by Western and CEE scholars alike, seems to take categories of difference, such as 'Western' or 'Eastern European' for granted, without attempting a relational reading of how such difference is constructed in the first place and to what end.
>
> (Koobak & Marling 2014, p. 331)

The consequence, taking for example the West-East division without questioning it, is to naturalise this difference.

All of these thoughts sounded like warnings during my thesis. Rather than questioning my subject of study (the Serbo-Croat border in the context of pre-accession to the EU), these thoughts pushed me to be vigilant about how to conceptualise spatially and temporally the way I was going to approach it. Among the epistemological and political stakes, it was notable to avoid observing the cross-border cooperation policy as if its establishment constituted, in my opinion, a proof of the democratisation or Europeanisation of formerly socialist territories—even if it were the intention of the EU. I focused instead on the potential social and spatial reconfigurations that this policy generates (or not) and reflects (or not) at the border. This critical posture finally led me to organise my field approach of the Serbia/Croatia border identified as the starting point and focus of attention, from which I was then able to observe the injunctions to cooperation and to reconciliation in the framework of pre-accession (the second object). To that extent, I have tried to position myself in rupture of the over-researched approach of (South-)East Europe through post-socialism by proposing instead to study how Europeanisation is perceived, re-appropriated, discussed, and manipulated from the Serbia/Croatia border and by considering the latter as a place-subject from which I think and not as an object on which I think.

Thinking post-Yugoslav, an illusory panacea?

Not reproducing the analytical frameworks of methodological nationalism or of post-socialism does not mean that it is a question of denying the importance of integrating nationalisms and the Yugoslav socialist past into the understanding of the Serbo-Croatian border. This critical positioning consists rather in refusing for oneself, in the elaboration of the field survey, the reifications of societies, spaces, individuals, and situations that the exclusive use of one or the other of the conceptual structures would entail. In this way, the goal is also to help update how these categorisations continue to be used prominently in social research conducted in the region. The question then becomes: how does one think outside of—even against—these exclusive frames of reference? Chari Sharad and Katherine Verdery invite, for example, a kind of intersectionality between the 'posts': 'we ought to think between the posts because they can offer complementary tools to rethink contemporary imperialism' (Sharad & Verdery 2009, p. 12).

Following this logic, I elaborated with two colleagues in a collective chapter published in 2015 on what seemed to us to be, then, the main strengths of the post-Yugoslav paradigm:

> in the plural and non-exclusive way in which it is defined here, the post-Yugoslav term escapes, in part, some of the normative limits pointed out in relation to other post-. It serves to translate social hybridity rather than dichotomy, synchrony rather than diachrony. It is conceived in contrast to nationalism to qualify (1) a voluntarily vague moment, that of the time after the dislocation of the political entity called Yugoslavia (postulating the survival and fluidity of certain ideas related to it); (2) a space defined by unclear human practices, territories and societies with often common histories and representations, whose proximities and socio-spatial exchanges sometimes persist, change and redeploy. It aims to translate persistences and resistances, not just breaks, without postulating the rails of a linear progression or a homogeneous spatial dispersion, but, on the contrary, of simultaneities and divergences, without implying either an objective, a model, necessarily better, supposedly more democratic.
>
> (Blondel et al. 2015, pp. 14–15)

Because we used the term post-Yugoslav in the special issue we were coordinating, it seems to me that we contented ourselves with justifying its interest, that is to say, above all to highlight the advantages of using this category. However, as Andreas Wimmer and Nina Glick Schiller point out, any conceptualisation leads to limiting and formatting one's perspective to neglecting certain elements at the expense of others to which one then pays an exaggerated attention (Wimmer & Glick Schiller 2003).

Thus, what constitutes the main advantage of the 'post-Yugoslav' category of analysis is perhaps also its main limitation. Conceived in response to the nationalisms of the 1990s, the term remains based on the historical essentialisation of a single temporal break (the collapse of Yugoslavia as zero time) and may appear as anchored in the nostalgia for an idealised political and societal project (the 'third way' of Yugoslavian self-managing socialism). By wielding a political and territorial particularism as a basis for understanding contemporary phenomena that, above all, would be understood only as specifically regional, it does not really make it possible to go beyond the Balkan aporias pointed out over several decades (Bakić-Hayden 1995; Todorova 1997). Finally, one may wonder whether to use the term post-Yugoslav does not lead to falling into the three traps of methodological nationalism at the same time, simply by moving them to another scale. Does this category of analysis not risk enclosing the researcher themself in the *Yugonostalgia* they claim to capture? Conceived as a way of escaping an approach centred on the nationalisms of the 1990s (at the forefront of which were Serbian and Croatian), does it not risk leading to focus too much on (and overvaluing) the Yugoslav legacy?

Here, I do not postulate that post-Yugoslav scale and spatiality are over-researched *per se*. I rather hypothesise that Yugonostalgia may be both an emotion and a political statement that are under-problematised. For researchers from outside the region, in particular, it may be projected as an awkward—yet still Western in essence—way to establish a link with a little-known territory and population that one wishes to investigate. For all, post-Yugoslav overthinking may constitute a blind spot when this position becomes an anti-nationalist gimmick forming nothing but its decalcomania, its negative, and then its shadow. The post-Yugoslav imaginary, which we can doubt that it corresponds more to reality than the nationalist representation, acts as a counter-ideology instead of being repositioned as one of the possible imaginaries, being the object of struggles in which its opponents and supporters participate.

How, then, can we go beyond this limit? In this case, the comparative intersectionality of Sharad and Verdery appears to be insufficient. Juxtaposing the conceptual structures of post-communism and postcolonial or postimperial theory does not allow us to go beyond the blind spots common to all these approaches. What is necessary for the opening of a real dialogue between the approaches is to make the intersectionality of hermeneutics the starting point of the research: 'Instead of comparing everything and everyone with the Western ideal used as a model for the whole of humanity, we can turn to an imperative mutual learning process based on pluritopic hermeneutics' (Tlostanova 2010, p. 131). This means both escaping universalist applications of ready-to-use discourses and travelling theories to start from the diversity of subjectivities and experiences of local histories marked by colonial and imperial differences (or their combination) within modernity/coloniality (*ibid.*).

Tlostanova's conclusion on the former Soviet space can then inspire that of this section on the post-Yugoslav space. The post-socialist, postimperialist, and postconflict connotations intersect and communicate constantly in the complex imaginary of the post-Yugoslav space, leading to the nostalgia and recycling of imperial and nationalist myths. What seems ultimately necessary is what might be called 'de-Yugoslavisation'. Returning to the previous model does not allow us to go beyond the dichotomy of Yugoslavia/nationalisms since both are constructed in opposition, thus mirroring one another. The 'de-Yugoslavisation' refers to a new impetus just like 'de-Sovietisation':

> Such an impulse is based not on negation or self-victimization, nor on violence, but on the creation of something different, other than modern/ colonial/socialist, taking its own path, superseding the contradictions inherent in these categories. In this context, creolization, hybridity, bilingualism, the psychology of the

> returned gaze and the colonialist/ colonizer intersection, as well as a stress on transculturation instead of acculturation and assimilation, can already be found in their specific postsocialist forms, which often parallel postcolonial ones.
> (Tlostanova 2012, p. 138)

This last point needs to be clarified because the stakes are multiple. On the one hand, it is a question of pointing to the relationship of domination established by the discursive use of stereotypes of Balkanity, in the post-Yugoslav context, from the outside by the West, which tries to impose its modernity. On the other hand, it is also a matter of deconstructing the use of Balkanism, within the region and on different scales, as an instrument of territorial and social differentiation (Obad 2014). Nevertheless, it is not a question of essentialising the regional scale in itself by singling out the post-Yugoslav experience. In other words, Balkanism is a particular type of Orientalism but remains Orientalism (observable in other Eastern European and more generally post-Soviet spaces) (Todorova 2011). Understanding the flexibility but also the circumstances of the use of representations of *the Other* in the process of differentiation and European integration becomes a central issue:

> This reinscription of otherness . . . functions not as a clear-cut binary but as a more flexible and contingent attribution of Europeannness versus Eastness to different places. It operates through multiple demarcations, which share the opposition of Europe and the East but delineate these categories differently.
> (Kuus 2004, p. 484)

This has implications for the researcher approaching the socio-spatial reconfigurations by integrating the geopolitical categories of knowledge, allowing the historical recontextualisation of their uses as much as the criticism of the presuppositions of exceptionalism on which they rest. This seems to me one of the contributions of the decolonial option.

The decolonial option: repoliticising the ethnographic situation

Conceiving inheritances as indissociably colonial and modern, the decolonial option articulates economic, sociological, and historical analyses with philosophical developments (Boidin 2010, pp. 131–132). Culture is then thought of as a constitutive of capitalist accumulation processes (*ibid.*).

Faced with the limitations of using the paradigms previously discussed, some researchers, primarily from the *Souths* (Castro Gómez & Grosfoguel 2007; Escobar 1995; Mignolo & Tlostanova 2006; Mignolo 2007), propose an epistemic break to deconstruct the discursive bases of the modernist and colonial project (the one and the other going together according to them), and thus to expose the coloniality of knowledge:

> Coloniality of knowledge is a typically modern syndrome, consisting of all models of cognition and thinking, and interpreting the world and the people,

> the subject-object relations, the organisation of disciplinary divisions, entirely dependent on the norms and rules created and imposed by western modernity since the 16th century, and offered to humankind as universal, delocalised and disembodied.
>
> (Tlostanova 2015, p. 39)

According to this perspective, modernity in itself is not an objective historical process, it is above all a system generating the hegemonic narrative of Western civilisation. Some aspects of the story are told in a certain way and are presented as an objective ontological reality. And the knowledge system on which this story rests becomes an instrument for disavowing other forms of knowledge, pushing them away from modernity:

> The co-existence of diverse ways of producing and transmitting knowledge is eliminated because now all forms of human knowledge are ordered on an epistemological scale from the traditional to the modern, from barbarism to civilization, from the community to the individual, from the orient to occident.
>
> (Castro-Gómez 2007, p. 433)

Scientific thought is then positioned as the only valid form of knowledge production. And Europe thus acquires an epistemological hegemony over all the other cultures of the World, which leads the researcher to a 'zero-point hubris' (*ibid.*). Tlostanova describes the latter as an arrogant desire to take the position of the outside observer (which thus cannot be observed), supposedly cleared of any bias or subjective interest claiming to seek pure truth and not compromised (Tlostanova 2015, p. 41). Both territorial and imperial, this epistemology is based on 'doctrines of theological (Renaissance) and egological (Enlightenment) knowledge based on the suppression of sensitivity, of the body and of its geo-historical deeprootedness that enabled them to claim it as universal' (Mignolo 2013, p. 183).

The theoreticians of the decolonial option invite us to change the biography and geography of reason by accepting the plurality of geo-doctrines of knowledge and the plurality of corpo-doctrines of feeling, believing, and understanding. To achieve this, decolonial thinkers advocate a border epistemology that focuses on 'changing the terms of the discussion and not just on its content', which means disengagement from Western democracy, capitalism, and communism as the only ways of thinking, doing, or living (Mignolo 2013, p. 182). Enacting border thinking requires a *sensitivity to the world* that is not a *view on the world* because this favourite expression of Western epistemology blocks feelings and the sensory fields beyond vision (*ibid.*). In this sense, it constitutes an act of epistemological disobedience: thinking and acting in a decolonial way comes about by inhabiting and thinking of the borders of local histories confronted with global designs. Its purpose is to demonstrate that modernity (peripheral or not, subordinate or not, alternative or not) is also only an option and not the 'natural' course of time (*ibid.*).

Beyond the arguments it provides to the critique of classical analytic categories, what can the decolonial option bring to the epistemological reflection conducted in Europe? What can the reading of thoughts developed first in the Latin American context produce? And would not trying to transfer them be to betray their epistemological and ontological roots in the experiences and struggles of the *Souths*?

As Capucine Boidin points out, decolonial and postcolonial studies provoke debate and resistance in the French social sciences. She notes three main criticisms: 'United States-centrism, Manichaeism and essentialism' often united in a 'capital sin: communitarianism' (Boidin 2010, p. 129). But one may wonder: is not this opposition in principle used to clear oneself in advance of the responsibility for an in-depth interpretation? Everything happens as if it were difficult to conceive of insights from traditions that are considered peripheral could bring relevant perspectives to the world. To exclude them from the realm of knowledge and from the academic agenda is to consider them as objects of knowledge, and not as creators of knowledge, or else as 'necessarily local knowledge with a local scope' (*ibid.*). That would be to discredit them by invoking precisely the reason why the authors have elaborated such a thought.

Contrary to this conservative reaction, I chose to enlist the decolonial approach in my thesis on the Serbia/Croatia border. In this way, I joined the ongoing work that has recently transposed this analysis, conducted first in the American context, to the second world (East Europe, Balkans, Caucasus, and post-Soviet space). They start from the same observation. To not reduce the superposition and the complex rivalry between different forms of epistemic colonialism that run through discourses and imagery, it is a question of rejecting the rhetoric of modernity and its reductive simplification—the opposition between the modern (Western by default) and the traditional (which requires the approval of a neocolonial power). In this way, the change in the discourse of modernity becomes more apparent as, for example, in the post-Soviet context: 'today the formula *national in its form, socialist in its content* gives way to a different one: *market and developmentalist in its essence, official-ersatz-ethnic-national in its form*' (Tlostanova 2012, pp. 138–139).

A better understanding of these often-forgotten spaces after the cold war requires taking into account multiple and successive wounds: 'the problematic of subaltern empires (Austria-Hungary, the Ottoman Sultanate, Russia) which act as intellectual and mental colonies of the first-rate capitalist Western empires in modernity, and consequently, create their own type of secondary colonial difference" (Tlostanova 2009, p. 4). In the case of the field chosen in my thesis, integrating the Austro-Hungarian and Ottoman secondary colonial differences allowed me to better apprehend the discursive and reflective use of the Balkans as an 'incomplete other of Europe' (Todorova 1997). Decolonial thinking has also allowed me to better understand the entanglement of successive colonial projects. In fact, on the aforementioned Austro-Hungarian and Ottoman base frame, another modernity, socialist, 'mutant, marginal, yet resolutely Western in its way of thinking and acting, has been printed, a global emancipatory utopia that has become reactionary and conservative' (Tlostatova 2009, p. 4).

The third aspect (and interest) in the application of decolonial thinking to South-Eastern Europe is the reflection on the syndrome of self-colonisation (especially researchers). According to Tlostanova, this is the most difficult element to apprehend (and to overcome) in the North-West domination of Southeast Europe, but also the most crucial:

> Within the world of imperial difference all modernity discourses acquire secondary, othered and mutant forms. This refers to secondary Eurocentrism practiced by people who have often no claims to it . . ., to secondary Orientalism and racism that flourish particularly in relation to the non-European colonies of subaltern empires . . . giving them a multiply colonized status and a specific subjectivity often marked with self-racialisation and self-orientalising. Without these additional categories we cannot rethink humanities, social movements or subjectivities in these spaces, we cannot hope to de-colonize or de-imperialize them.
>
> (*ibid.*)

The decolonisation of knowledge then requires first and foremost the decolonisation of the research produced on this space-time. The main difficulty probably lies in the negation by cultural producers of 'multiple subjectivities, distorted reflections' typical of the 'Second world' (*ibid.*). But these do not correspond to what one finds in 'the enormous supermarket of ideas, thoughts, theories, philosophies, religions proposed by the modern world' (Kaplinski 2002). This diversity does not seem to correspond either to the Western approach of scientific thinking as the only valid form of knowledge production (Castro-Gómez 2007). This last point has constituted (and still constitutes today) a source of inspiration, decolonial thought as the thread of an epistemological, reflexive, and critical gymnastics for research in Europe and on Europe, like that which can be conducted elsewhere and on elsewhere.

Thus, more than a positioning, the decolonial option proposes a political agenda, undoubtedly idealistic, for independent research and at the same time more understanding and more critical, in particular, on the post-Yugoslav space-time, which feeds universities that emancipate individuals:

> The value of any independent social approaches then would be linked with their ability to . . . turn to the goals and tasks of academia that have been long forgotten, such as the crucial aim of the university to shape not a submissive and loyal narrow specialist in some applied science but first of all a critically thinking self-reflexive and independent individual, never accepting any ready-made truths at face value, truly and unselfishly interested in the world around in all its diversity and striving to make this world more harmonious and fair for everyone and not only for particular privileged groups. And is this not ultimately the true mission of a vigorous decolonized social theory?
>
> (Tlostanova 2015, p. 54)

Breaking the stereotypical way in which former Yugoslavia has been over-researched requires learning to unlearn in order to relearn other bases and frames of thought and sometimes to create new thoughts or reshape existing ones. Thus, and this constitutes an underlying assumption of this chapter, the decolonial option is a political and epistemological choice, which leads to taking more into account the historical and cultural balance of power in the elaboration and understanding of the ethnographic situation. This is important for the investigator and the respondent. The risk may be, behind the imperative of the struggle against cultural essentialism, to print a kind of political essentialisation, which would amount to reading every act and every word (including their absence), every observation and every field report, and more broadly any research situation, from the exclusive angle of its political sense.

Conclusion

This chapter was an opportunity to review and discuss the epistemological and political questions posed by the research conducted in my thesis. It reflects my awareness of certain elements of the geopolitical context (the enlargement of the EU as a modernist absorption project) as well as its latent impregnation in the way the question is asked by most researchers, including me (the reproduction of the modernist reading grid to observe this phenomenon). The majority of those involved in setting up the European project, but also in its analysis, are affected by the colonisation or self-colonisation 'syndrome'. Regardless of the places in which they act and of which they speak—which could also be described as the interpenetration of scales of domination—they offer a dichotomous reading of the phenomenon and the way in which it is studied. The demarcation of modernity and its discourses (in an epistemic sense) make it necessary to have a greater reflexivity on the concepts and the methods used. This is to *at least* make apparent the unavoidable biases of the situation ethnographic, as well as trying to limit its participation in the maintenance of dominant discourses through the use of its grammar. The researcher's awareness of his own limitations also means reflecting the social status they embody in the field. To stand out from modernity is not only about oneself but also about others' perceptions of themselves. Without it being possible to reach the truth of the social world:

> In a certain way, as far as the social world is concerned, the perspectivism as defined by Nietzsche cannot be surpassed: everyone has his truth, everyone has the truth of his interests . . . If there is a truth, it is that this truth is the object of a fight.
>
> (Bourdieu 2015)

This chapter is also the story of my research disenchantment, which is probably an inevitable contingency of the thesis exercise. The revelation of the difficulties and ambiguities of organising and carrying out fieldwork highlights what Didier Fassin (2008b, p. 13) describes as an ethnographic test, 'a risk-taking that begins in the inquiry relationship and extends into the work of writing . . . beyond the singularity of experiences'. But as he points out: 'These issues concern nothing less than the conditions of the verification of the investigation, the human relationship in which it is anchored, the results we can draw from it and the social effects that we produce in doing so' (*ibid.*, p. 14). Through these reflections, I interrogate (and invite those who wish to do so with me) my/our political and social responsibilities through the choice of words and concepts (nationalism, post-socialism, and post-Yugoslav) I use when I over-research certain objects and certain places asking the same questions—often to facilitate our inter-understanding and in the interest of our careers rather than to increase the understanding of a phenomenon or a situation.

Thus, in my interpretation of over-research of and in former Yugoslavia, I have tried to point out the exhaustion of the study of peace, stability, Europeanisation, memory, war and post-conflict, nationalism, instability, and post-socialist transition (in particular), since all these themes seem to be tinged with a *Western* a priori and therefore with a modern/colonial pre-supposition. It is not a question of saying here that every research on former Yugoslavia (and similarly on Cyprus, or Northern Ireland or other post-conflict territories of *Western* obsession) is constructed from this angle but that it remains the dominant angle from which the region and its populations are thought, approached, and described by cultural producers. And I am talking here as much about numerical domination as symbolic domination. For if other works exist, they often remain marginalised. As if the banal and common questions of our time, the daily life of change, were less important (e.g. the questions around European solidarity and migration, but also activism and the fight against discriminations; see Bilić & Kajinić 2016).

In this context, over-research leads to erosion and weakening of what we produce in a tautological way as much as of those we investigate, more and more tired of being the *object* of our pathetic curiosity. It also emphasises the impossible emergence of critical geopolitics in the production of knowledge. Our obsession with the imagined savagery of the other finally says more about ourselves than about the territories and populations that are the object of our fantasies. And always this difficulty of truly opening ourselves to otherness is translated in our approaches by our incapacity to grant the peripheral populations the status of subject. In other words, research is also about responsibility: ‘The link with others is only established as a responsibility, whether or not one accepts it, whether or not one knows how to assume it, whether or not one can do something concrete for others’ (Levinas 1982, p. 93).

Notes

1 This chapter is an English translation and adaptation of a paper initially published in French: Blondel, Cyril, « Gymnastique épistémologique, critique et réflexive: la construction d’un terrain de recherche en ex-Yougoslavie face à la colonialité du savoir ». *Nouvelles Perspectives en Sciences Sociales*, Vol. 13, No 1., Déc 2017. pp. 57–89.

2 All translations from French are by the author.

References

Agnew, John A. (2014) “Le piège territorial. Les présupposés géographiques de la théorie des relations internationales”, *Raisons politiques*, vol. 2, no. 54, pp. 23–51.

Bakić-Hayden, Milica (1995) “Nesting orientalisms: The case of former Yugoslavia”, *Slavic Review*, vol. 54, no. 4, pp. 917–931.

Beck, Ulrich (2003) *Pouvoir et contre-pouvoir à l’ère de la mondialisation*, Paris, Flammarion.

Bilić, Bojan and Sanja Kajinić (2016) *LGBT Activist Politics and Intersectionality: Multiple Others in Serbia and Croatia*, London, Palgrave Macmillan.

Blondel, Cyril (2013) "La coopération transfrontalière, un levier potentiel des réconciliations interethniques en ex-Yougoslavie? Une approche critique", *Cybergeo: European Journal of Geography*, vol. 641, https://journals.openedition.org/cybergeo/25881

Blondel, Cyril (2016) *Aménager les frontières des périphéries européennes: la frontière Serbie/Croatie à l'épreuve des injonctions à la coopération et à la réconciliation*, doctoral thesis, Tours, Université François Rabelais.

Blondel, Cyril (2021) "How approaching peripheralisation without peripheralising? Decolonising (our) discourses on socio-spatial polarisation in Europe", *Justice Spatiale | Spatial Justice*, no. 17.

Blondel, Cyril, Guillaume Javourez and Marie van Effenterre (2015) "Avantpropos. Habiter l'espace post-yougoslave", *Revue d'études comparatives Est-Ouest*, vol. 46, no. 4, pp. 7–34.

Boatcă, Manuela (2006) "Semiperipheries in the world-system: Reflecting Eastern European and Latin American experiences", *Journal of Worldsystems Research*, vol. XII, no. II, pp. 321–346.

Boatcă, Manuela et Sérgio Costa (2010) "Postcolonial sociology: A research agenda", in Manuela Boatcă, Sérgio Costa et Encarnación Gutiérrez Rodríguez (eds.), *Decolonizing European Sociology*, London, Ashgate, pp. 13–31.

Boidin, Capucine (2010) "Études décoloniales et postcoloniales dans les débats français", *Cahiers des Amériques latines*, vol. 62, pp. 129–140.

Bourdieu, Pierre (1991) *Language and Symbolic Power*, Cambridge, Harvard University Press.

Bourdieu, Pierre (2015) *Sociologie générale: Cours au Collège de France 1981–1983*, Paris, Seuil.

Brubaker, Rogers (2002) "Ethnicity without groups", *European Journal of Sociology*, vol. 4, no. 2, pp. 163–169.

Castro-Gómez, Santiago (2007) "The missing chapter of empire: Postmodern reorganization of coloniality and post-Fordist capitalism", *Cultural Studies*, vol. 21, nos. 2–3, pp. 428–448.

Castro-Gómez, Santiago and Ramón Grosfoguel (2007) *El giro decolonial, Reflexiones para una diversidad epistémica mas allá del capitalismo global*, Bogota, Siglo del Hombre Ed.

Chiclet, Christophe (1997) "Transition démocratique dans l'ex-Yougoslavie", *Confluences Méditerranée*, no. 2, pp. 103–109.

Dhorliac, Renaud (2014) "Vingt ans d'ex-Yougoslavie: une transition générationnelle inachevée", *Annuaire français de relations internationales*, vol. XV, pp. 133–149.

Dumitru, Speranta (2014) "Qu'est-ce que le nationalisme méthodologique? Essai de typologie", *Raisons politiques*, vol. 2, no. 54, pp. 9–22.

Erdei, Ildiko (2009) "Hopes and visions. Business, culture and capacity for imagining local future in Southeast Serbia", *Etnoantropološki problemi*, vol. 4, no. 3, pp. 81–102.

Escobar, Arturo (1995) *Encountering Development: The Making and Unmaking of the Third World*, New York, Princeton University Press.

Fassin, Didier (2008a) "Répondre à sa recherche. L'anthropologue face à ses "autres"", in Didier Fassin and Alban Bensa (dir.), *Les politiques de l'enquête*, Paris, La Découverte, pp. 299–320.

Fassin, Didier (2008b) "Introduction. L'inquiétude ethnographique", in Didier Fassin and Alban Bensa (dir.), *Les politiques de l'enquête*, Paris, La Découverte, pp. 7–15.

Ferguson, James and Akhil Gupta (2002) "Spatializing states: Toward an ethnography of neoliberal governmentality", *American Ethnologist*, vol. 29, no. 4, pp. 981–1002.

Gaillard, Edith (2013) *Habiter autrement: des squats féministes en France et en Allemagne. Une remise en question de l'ordre social*, doctoral thesis Tours, Université François Rabelais.
Hadjimichalis, Costis and Ray Hudson (2007) "Rethinking local and regional development: Implications for radical political practice in Europe", *European Urban and Regional Studies*, vol. 14, no. 2, pp. 99–113.
Jansen, Stef (2015) *Yearnings in the Meantime: "Normal Lives" and the State in a Sarajevo Apartment Complex*, Oxford; New York, Berghahn Books.
Kaplinski, Jaan (2002) *The Brave New Merry-go-round*, http://jaan.kaplinski.com/opinions/merry-go-round.html, accessed 19 September 2017.
Koobak, Redi and Raili Marling (2014) "The decolonial challenge: Framing post-socialist Central and Eastern Europe within transnational feminist studies", *European Journal of Women's Studies*, vol. 21, no. 4, pp. 330–343.
Kuus, Merje (2004) "Europe's eastern expansion and there inscription of otherness in East-Central Europe", *Progress in Human Geography*, vol. 28, no. 4, pp. 472–489.
Leca, Jean (2000) "Sur la gouvernance démocratique: entre théorie et méthode de recherche empirique", *Politique européenne*, vol. 1, no. 1, pp. 108–129.
Levinas, Emmanuel (1982) *Ethique et infini*, Paris, Fayard.
Massey, Doreen (1999) "Imagining globalization: Power-geometries of timespace", in Avtar Brah, Mary Hickman et Mairtin Mac an Ghaille (eds.), *Global Futures: Migration, Environment, and Globalization*, New York, St Martin's Press, pp. 27–44.
Mignolo, Walter D. (2007) "Delinking. The rhetoric of modernity, the logic of coloniality and the grammar of de-coloniality", *Cultural Studies*, vol. 21, nos. 2–3, pp. 449–514.
Mignolo, Walter D. (2013) "Géopolitique de la sensibilité et du savoir. (Dé)colonialité, pensée frontalière et désobéissance épistémologique", *Mouvements*, vol. 1, no. 73, pp. 181–190.
Mignolo, Walter D. et Madina Tlostanova (2006) "Theorizing from the borders. Shifting to geo- and body-politics of knowledge", *European Journal of Social Theory*, vol. 9, no. 2, pp. 205–221.
Neveu, Catherine (2007) "Introduction", in Catherine Neveu (dir.), *Cultures et pratiques participatives. Perpectives comparatives*, Paris, L'Harmattan, pp. 13–30.
Obad, Orlanda (2014) "On the privilege of the peripheral point of view: A beginner's guide to the study and practice of Balkanism", in Tanja Petrović (ed.), *Mirroring Europe. Ideas of Europe and Europeanization in Balkan Societies*, Leiden, Koninklijke Brill, pp. 20–39.
Pasquier, Romain (2012) "Comparer les espaces régionaux: stratégie de recherche et mise à distance du nationalisme méthodologique", *Revue internationale de politique comparée*, vol. 19, no. 2, pp. 57–78.
Petrović, Tanja (2014) "Introduction: Europeanization and the Balkans", in Tanja Petrović (ed.), *Mirroring Europe. Ideas of Europe and Europeanization in Balkan Societies*, Leiden, Koninklijke Brill, pp. 3–19.
Sarkar, Mahua (2004) "Looking for feminism", *Gender and History*, vol. 16, no. 2, pp. 318–333.
Scholte, Jan Aart (2000) *Globalization: A Critical Introduction*, New York, Palgrave Macmillan.
Sharad, Chari and Katherine Verdery (2009) "Thinking between the posts: Postcolonialism, postsocialism, and ethnography after the Cold War", *Comparative Studies in Society and History*, vol. 51, no. 1, pp. 6–34.

Štiks, Igor (2009) "L'européanisation des pays successeurs de l'ex-Yougoslavie: la fin de la conception ethnocentrique de la citoyenneté", in *L'Europe sous tensions. Appropriation et contestation de l'intégration européenne*, Paris, L'Harmattan, pp. 281–304.

Suchland, Jennifer (2011) "Is Postsocialism Transnational?", *Signs*, vol. 36, no. 4, pp. 837–862.

Tlostanova, Madina (2009) *Towards a Decolonization of Thinking and Knowledge: A Few Reflections from the World of Imperial Difference*, https://antville.org/static/sites/m1/files/madina_tlostanova_decolonia_thinking.pdf, accessed 19 September 2009.

Tlostanova, Madina (2010) *Gender Epistemologies and Eurasian Borderlands*, New York, Palgrave Macmillan.

Tlostanova, Madina (2012) "Postsocialist ≠ postcolonial? On post-Soviet imaginary and global coloniality", *Journal of Postcolonial Writing*, vol. 48, no. 2, pp. 130–142.

Tlostanova, Madina (2015) "Can the post-soviet think? On coloniality of knowledge, external imperial and double colonial difference", Intersections. *East European Journal of Society and Politics*, vol. 1, no. 2, pp. 38–58.

Todorova, Maria (1997) *Imagining the Balkans*, New York; Oxford, Oxford University Press.

Todorova, Maria (2011) "Balkanism and postcolonialism, or on the beauty of the airplane view", in Costică Brădăţan and Sergej Aleksandrovič Ušakin (eds.), *In Marx's shadow: Knowledge, power, and intellectuals in Eastern Europe and Russia*, Lanham; Boulder; New York, Lexington Books, pp. 175–195.

Wacquant, Loic (2007) "La stigmatisation territoriale à l'age de la marginalité avancée", *Fermentum*, no. 48, pp. 15–29.

Wacquant, Loïc, Tom Slater and Virgílio Borges Pereira (2014) "Territorial stigmatization in action", *Environment and Planning A*, vol. 46.

Wallerstein, Immanuel (2014) *The Modern World-System*, New York, Academic Press.

Wimmer, Andreas et Nina Glick Schiller (2003) "Methodological nationalism, the social sciences, and the study of migration: An essay in historical epistemology", *International Migration Review*, vol. 37, no. 3, pp. 576–610.

Žižek, Slavoj (1994) "The spectre of ideology", in Slavoj Žižek (ed.), *Mapping Ideology*, London; New York, Verso, pp. 1–33.

4 Ghosts of researchers past, present, and future in Mumbai

Cat Button

Mumbai's ghosts

We are all ghosts to other researchers, and sometimes we even haunt ourselves. This chapter suggests a reflexive approach to research that acknowledges the effects of researchers that have undergone before and the impact on those that will follow. Certain places attract high concentrations of research and Mumbai is one such place with many past, present, and future researchers. The seed of this chapter was sown whilst watching the Bollywood film *Dhobi Ghats: Mumbai Diaries* (Rao, 2011) at a cinema in Mumbai during fieldwork. The story of the movie follows the intertwining experiences of four different characters to give a rich insight into lives in Mumbai. One character sparked the thought process that culminated in this chapter: Shai is young woman who has spent most of her life in the United States of America and has now returned to India on a sabbatical to research traditional livelihoods (*ibid.*). Bollywood films have documented life in Mumbai for decades (Gangar, 2003; Pendse, 2003; Prakash, 2010) and this acceptance of a researcher as a main character indicates that encounters with research practices are now a part of everyday life in Mumbai. This led to reflection on my own presence and positionality as a researcher to further consider the consequences of acts of research here.

Mumbai is not unique as a city where being researched has become an everyday experience. However, Sukarieh and Tannock (2013: 494) note in their research on the Shatila Palestinian Refugee Camp that 'over-research remains under-addressed by social scientists' and is in need of further consideration. As focus moved away from the narrow gaze at North American and European cities, Urban Studies made a call, most significantly by Jennifer Robinson (2006), to not repeat the pattern of focusing on large cities seen to be different and to instead look at the 'ordinary city'. This led the turn in postcolonial urban studies towards thinking of each city as 'ordinary' and worthy of study, particularly in response to the use of cities in the Global North as case studies for the understanding of all urban conditions. Yet, the idea that 'paradigmatic cities' (see Nijman, 2000) might inform us about the current zeitgeist generally continues in the Global South, with the fetishisation of megacities or uniquely interesting metropolises such as Mumbai. These cities are now being intensively researched and used to describe other urban areas in their respective region and beyond, thus replicating many

DOI: 10.4324/9781003099291-5

of the issues that the new focus was striving to move away from. Certain places continue to attract such a huge number of researchers that mean negotiating, not just encounters with, the traces of previous researchers and also with contemporaries (see Neal et al., 2016). That there are particular cities where research activities are so commonplace suggests a need for more reflection on shared research practices, epistemologies, and the urban theory that emerges from these contexts. Perhaps of even more significance is how concepts derived from heavily researched sites are then extrapolated to also speak for other places. The intensive and incessant research of certain cities is thus problematic to the understanding and conceptualisation of the urban by limiting the experiences of researchers and also altering those experiences in a feed-back loop.

This chapter describes encounters with other researchers (past, present, and future) and considers the effect on both respondents and data. The issues of high numbers of researchers in one place are explored through the reflection of empirical work on Mumbai. First, I use literature and my experiences to explore how past researchers haunt current research activities in the traces they leave in researcher behaviour and respondents' expectations. Secondly, I use empirical evidence from fieldwork in Mumbai to describe encounters with contemporary researchers. Other researchers are encountered in the field, often in an ad hoc manner, and this presents opportunities and challenges for everyone involved. Vignettes from this research are presented to explore how the researcher ghosts haunting Mumbai might affect data collection and analysis, and to imagine the effect this must also have on other researchers from all disciplines and industries. In this chapter, I focus on the micro-scale to consider what effects the number of researchers is having on the individual research project, using an example from my collection of empirical data on middle-class housing and rainwater harvesting (see Button 2017). Thirdly, I consider the presence of future researchers to develop discussions of positionality in research to encompass inter-researcher experiences. The potential future researcher becomes folded into the present as the audience for current activities.

The implication is not that research-intensive urban spaces should be considered too tarnished and that urban researchers should seek cities heretofore untouched by researcher activities. Rather, these ghosts are used to provoke reflexivity and recognition of the positionality of not just as oneself but as one of the host of researchers (past, present, and future) haunting the fabric of certain cities. I am looking at positionality in a range of relationships. The positionality between a researcher and a respondent is often one sided; but, here I ask one to consider the relationship between researchers and also between research ideas. We are all in flux and cycle between the positions of past, present, and future researchers and this chapter calls for the acknowledgement of these complexities in our approaches to proposing, undertaking, and presenting our research.

Ghosts of past researchers

The traces of previous researchers are seen in the literature and their echoes were felt during fieldwork whilst researching middle-class housing and environmental initiatives, particularly rainwater harvesting. Encounters with past researchers can change the behaviour of respondents, making the process easier, harder, or just different. The number and frequency of researchers in Mumbai can have several effects on participants. In this section, I use literature to discuss how previous researchers cause research fatigue, change respondent expectations, and then (unintentionally) encourage people to develop coping mechanisms to deal with being researched.

'Research fatigue' of participants is something that I encountered in Mumbai, especially when interviewing government officials during my nine months of fieldwork (see Button, 2014). Research fatigue is often considered in the context of long-ongoing studies and participants dropping out but it is also encountered when the same person or group is researched repeatedly by different people (Boesten and Henry, 2018) or even by several researchers at the same time (Koen et al., 2017). A lot of people's time is taken up in heavily researched communities and by a dominant interest in a small number of topics and activities (Clark, 2008). This can also lead to people questioning their identities as they or their communities are repeatedly singled out as case studies for research. Identities can thus emerge through the research process, and research can directly change the people or place (Rose, 1997). Some of my respondents manifested this fatigue as annoyance, particularly at the time that is consumed by interviews and being asked similar questions repeatedly. Neal et al. (2016) found some of the same issues with over-researched communities in Hackney (London) where people and groups had developed mechanisms for dealing with being constantly researched and resisted being the subjects of further research. This is particularly seen when no change or benefit is seen after research, and then similar projects follow. The problem of research fatigue is often treated as something to conquer or work around. Treating research fatigue in this way, as an obstacle to be overcome, leads to a lack of reflexive attention to how research practice is informed by prior activities and, in turn, what kinds of knowledge and understanding the research can generate. We should reflect on why research fatigue occurs and what we can do differently. Perhaps, change the site or make the research of explicit value to participants.

An issue arising from the overuse of certain informants is that they may have answered questions so much that their answers are rehearsed. When the same person is involved as a respondent in several research projects, then they can become frustrated with the process and even with the requests. This can make the respondent reluctant to participate and alter their attitude to the research (Clark, 2008). This leads to the issue of people assuming that research is on a certain topic and not necessarily fully listening to the explanation. Urban areas of the Global South are often characterised by their informal settlements and discourse about megacities eclipses all other urban enquiry, leading to the suggestion of slum as a theory (Rao, 2006; Davis, 2004). However, this concentration on slums and megacities has led some scholars to re-think the way we conceptualise cities in the Global South (Roy, 2011; Robinson, 2006).

The fieldwork for my research project was with the middle-classeses to collect empirical data on housing issues and there was some confusion as many researchers come to Mumbai to research informal housing. Assumptions about research topic are very frustrating when respondents have pre-prepared answers that did not relate to my questions. I had to explain several times that I am not researching informal settlements, but middle-class housing. Respondents would still continue to tell me about informal settlements, and I presume this is what other researchers had spoken to them about. People often hear what they expect, not what is really said. The researchers before me influenced the way that I was positioned by others and also created precedent for what research focus in Mumbai must be. It is challenging to investigate an under-researched topic in an over-researched city. I can perhaps say that it is Mumbai's informal settlements that are the over-researched place, leading to these assumptions about what and where I must be focussing on. In this way, the ghosts of past researchers were there in my interviews and this subconsciously feeds into my writing.

Hammett and Sporton (2012) found that the expectations of respondents in rural Kenya had been shaped by their previous encounters with researchers. Respondents expected the same treatment from each set of the researchers, in particular, payment for their time after one group of researchers set this precedent. This raises questions of ethics and power, with payment not only being an ethical move to recompense participants' time commitments but leaving traces that effect the parameters of future research, especially research conducted on smaller budgets. Precedents also shape the nature of the data collected and the attitude to being-researched of respondents who have been researched heavily. In turn, individuals and communities develop strategies to deal with this and in such adaptations, we find the traces of past researchers. They clearly demonstrate the haunting effect that the approach of one past researcher can have on future research. There is also an understanding that the researchers that have gone before have not only changed the power dynamics and expectations but also the acknowledgement of their previous presence has had a less-tangible effect on the respondents in the way they react to researchers and to the answers they might give. These indirect effects are applicable to anyone collecting data in highly researched areas. Researchers need to be aware of the traces of research on target communities and reflect on the echoes that they will leave themselves.

Sukarieh and Tannock (2013: 504) note that 'Researchers, moreover, focus on individuals within these groups who they see as particularly good subjects.' This means that the same people are interviewed time and time again. It also sets up the notion that there are 'good subjects' and that respondents should behave in a particular way. This is a political act that allows external researchers to shape behaviours (McDowell, 1992). It could be problematic for research and knowledge. On the other hand, could be seen as a vital coping mechanism by communities. Aside from annoyance and fatigue, individuals and communities who are often researched may develop mechanisms to find ways to maximise the benefits of being research subjects.

Capitalising on being researched can have several benefits for communities. It can raise the profile of issues and could lead to beneficial change, meaning that the community in a place may not see themselves as over-researched (Koen et al., 2017). This tactic often also used by elites and politicians when being interviewed, in the way they will make sure their point is heard (Morris, 2009) but might be viewed as criticism of the status quo when others try to steer the agenda. Communities excluded from research projects may want to be included and envy the often-researched (Omata, 2019). It should be acknowledged that most research projects do not lead to tangible change and impact for communities, leading to frustration, boredom, and fatigue (Cleary et al., 2016; Sullivan et al., 2001). There is also a risk to becoming seen and heard by policymakers and so ethics and safeguarding must be carefully considered. Raising awareness can be beneficial but it can also leave people exposed to detrimental change.

Communities and individuals can also use being researched for financial gain. If researchers pay respondents directly, then the research itself can become an income stream. This sets up a precedent of payment for further researchers to negotiate (Head, 2009) and can lead to respondents telling the researcher what they think they want to hear to ensure they get repeat 'trade' (Sukarieh and Tannock, 2013). Offering any kind of incentive changes the field for the researchers that follow and is an ethical dilemma, partly because people may feel that they cannot refuse (Singer and Kulka, 2002). Other investments could also be made into a community as part of research, such as technology installation or training. There can be increased profit for local businesses from the visiting researchers needing places to stay, food and drink, and transport. Thus, whereas we see a lot of drawbacks of being over-researched, we can also see the potential benefits for communities. The effect on the data collected could be dramatic if certain communities deliberately make themselves into honeypots for research. However, if a community or area is labelled 'over-researched', this could divert researchers away and reduce opportunities (Koen et al., 2017).

What is generally missing from all these debates are the implications of other researchers on our data and analysis on the subsequent conclusions and theories that emerge from these to influence further studies. If the same places and people are included in research studies over and over again, then this will limit the breadth of knowledge and will lead to theories developed from a narrow frame. Beyond this, if respondents are rehearsing answers, either from fatigue or wanting to be a 'good respondent', then that skews the data further. Past researchers create unintended consequences that haunt the researchers that come after them and will echo through knowledge and theory unseen.

Ghosts of present researchers

Mumbai has so many researchers that others were physically present during data collection. The ghosts of these researchers emerge in data analysis and writings. My research in Mumbai concerned rainwater harvesting and how it is governed, assembled, and used in middle-class apartments. Original planned methods involved asking residents to complete diaries of everyday consumption practices to elicit deeper and carefully considered responses and to use auto-photography. However, with the middle-classeses and professionals as respondents, it rapidly became apparent that these methods would not be possible. Mandel (2003) demonstrates that no matter how much thought about positionality and consideration of feminist research practices is undertaken, assumptions made can differ radically from the situations encountered in context. The power dynamics with residents and with officials was balanced differently than I expected, and there was a strong resistance to any method beyond questionnaires and interviews (see also Smith, 2006; Welch et al., 2002; Cochrane, 1998). Crang (2005: 229) notes that research in human geography has become 'methodologically more uniform', relying on interviews in particular. This demonstrates how current practices are haunted by previous research. I found that to be taken seriously as a researcher then the methods acceptable to people were limited to these interviews and questionnaires. More creative (and often time-consuming) methods were not considered research and it was impossible to get people to agree to them. In this way, the respondents shaped the methods. I interviewed 59 people and small groups, with interviews lasting between 15 minutes and 2 hours. The majority of those interviewed are middle-class residents at their homes and professionals involved in rainwater harvesting (architects, NGO officers, government officials, etc.) in their offices. The research of material objects of infrastructure and buildings was primarily undertaken through site visits using photographs and videos.

Interviews with professionals were planned as semi-structured, with questions framed for each individual. These interviews usually became unstructured as the expert expanded on their views and experiences. I took the decision to let respondents take interview conversations in the direction that interested them and this often brought up new lines of enquiry. For example, in the interview with a developer (Mumbai, 16/12/2009), the conversation led to a discussion of the differences between everyday practices of living in India and North America. Another example is a Municipal Corporation of Greater Mumbai (MCGM) engineer (Mumbai, 10/02/2011) who took the conversation away from the municipality's interventions to explain that rainwater harvesting had now been installed in his own apartment building. These two examples demonstrate the strength of allowing respondents to uncover unexpected aspects and fresh avenues of enquiry. The main weakness of the approach is that the interview can stray into irrelevant topics. Having prepared questions and a clear aim of why this respondent is being interviewed helped to direct the conversation or steer it back.

Power and positionality are key issues in interviewing that can shape the way in which an interview is conducted and also the data collected (McDowell, 1992). Two main issues are the difficulties to access elite respondents and then how to control the interview (Rice, 2010). The literature on these two topics is divided broadly into those that discuss the reasons for conducting elite interviews and the power issues themselves (for example, Mandel, 2003; Ward and Jones, 1999) and 'how to' guides (such as Taylor et al., 2020). Morris (2009) points out that those in an elite position are likely to agree to interviewed when they have some point to get across and believe that talking to the researcher will advance their cause. If you are trying to contact someone with a lot of responsibilities, then they are likely to be short of time and wish to prioritise activities from which they can see a clear gain, in similar ways to the communities discussed earlier who deliberately position themselves as easy to research. Thus, professionals and elites refuse to be interviewed due to a lack of time for something of no value to them and there could be a lot of truth in this, with people only agreeing to interviews where they have something to convey and see the process as a way to communicate a message (*ibid.*). This may conversely mean some people are deliberately having themselves (over-)researched to get their point across in as many places as possible. Elites refusing to be interviewed can also be part of them not wanting to contradict the views of their company or to protect industry secrets (Ward and Jones, 1999). I found this with some professionals, for example, a prominent city planner in the municipality who presented the 'party line' of their position and said they did not know much about the topic and then opened up when I switched off the voice recorder and they felt they were no longer representing the establishment and could share more personal experiences. From these discussions, we can extrapolate that elites who are often interviewed for research have found coping mechanisms to either minimise time spent being researched or else to maximise potential publicity of their viewpoint. One of the consequences is that elite experts can find ways to reduce the amount of time they spend giving interviews by arranging meetings with different people at the same time, as discussed later in this section.

Contacting and meeting with people was difficult and this required flexibility in approach. A limited level of snowballing was achieved as some of the professionals would give me details of consultants or suppliers they worked with. Building up trust was a major factor and details were sometimes not passed on until after I had met the first person several times. This is perhaps due to previous research from competitors or marketing companies. We should be mindful that not all researcher ghosts are those of academics and not all research is held to the high ethical standards we expect from scholars. When asked if I could speak to someone else in the organisation, I was often told that I did not need to, as I had spoken to that 'one' person. At the other end of the spectrum, one interviewee tried to send me to speak to someone else instead, and I did eventually interview the second person as well. When respondents dismiss the importance of talking to others, this hints at ghosts of researchers past haunting the process making them wary of future involvement (Rice, 2010).

During this data collection, I unintentionally interacted with other researchers in Mumbai and then found quotes from people I have also interviewed presented at conferences. So, we see the power of elites to shape the encounter and thus the discourse. And also, how the discussion of power and positionality go much beyond the researcher–researched dynamic to consider the relationships and impacts between multiple researchers. Here, I will draw on examples to tease out the issues that arose and which could directly affect my data (as summarised in Table 4.1). First, unexpectedly sharing interviews with other researchers can alter the questions asked and the time you are given. Secondly, some officials have rehearsed answers to what they expect you to ask, having been interviewed so frequently.

The first encounter I had with another researcher was when I shared an interview slot with someone from a different discipline (occurrence 1 in Table 4.1). The interview was with the head of an NGO and had just begun when a social worker from Europe turned up to interview him as well. It was unclear whether this was an accidental double-booking or a deliberate action to save time by being

Table 4.1 Encounters with researchers in Mumbai, November 2009 to February 2011

	Occurrence	*Effect on data collection*	*Effect on analysis/ writing*
1	Social worker also interviewing same respondent	We were double booked. As topics were different it made it difficult to keep interview on my interests.	How to acknowledge questions posed by another researcher?
2	Postgraduate researcher already there	Allowed the interview to take place.	Was good for making contact with the society and I was able to visit again.
3	School children also interviewing same respondent	Only got to ask one question. Children distracted interviewee and they all spoke in Marathi. She also kept telling me to look at the website.	Limited my data collection. She did however pass me onto someone else in the local government.
4	Interviewing consultant whilst he was making TV documentary (because he was busy)	Was difficult to ask questions, as he was often on camera. I could not take notes easily or record sound.	Limited my collection of data but we did then visit a couple of other buildings as part of filming.
5	Marketing researchers also interviewing same respondent	I knew they would be there in advance and interviewed them too. Topic was similar and it was with someone I had interviewed before.	There is the issue of how to reference them (again).

interviewed by two people at once and thus take control (see Morris, 2009; Rice, 2010). Our research topics and objectives were very different which complicated the interview as it kept drifting off into discussions of social work and then was brought back to my focus on housing and the environment. It also meant gaining the permission from another party to be recorded and the interview lasted a lot longer than it would have if it had just been one researcher's questions. There are ethical dilemmas when it comes to using the answers to questions that were actually posed by another researcher who was also in the same interview spot. If I were to use an answer to one of the other researcher's questions, I would have to carefully consider the referencing and credit given. I have avoided this dilemma, as questions were usually not relevant to my research, except in occurrence 5 (see Table 4.1) where the interviewee (Co-operative Society Secretary for a middle-class residential building, Mumbai, 08/12/2010) gave his personal history of interest in the environment, which is something I had discussed with him before but not formally interviewed. I subsequently briefly interviewed the market researchers as well (Mumbai, 08/12/2010). The presence and positionality of other researchers in my interviews continues to haunt my writing a decade later.

One of the most complex encounters I had was whilst interviewing a Municipal Corporation of Greater Mumbai (MCGM) officer for rainwater harvesting who limited me to one question only due to lack of time (occurrence 3 in Table 4.1). She kept telling me to just look at the website. At the same time, she was patiently answering the questions (in Marathi) for the school project of an entire class of children who were also crammed into her office. This example shows clear evidence of research fatigue from the officer trying to do her job whilst being asked questions. And, I must admit, the school children were a more exciting audience than the foreign woman in the corner (me). So, the positionality between the respondent, the children, and me was important and complicated, with me having the least power in the dynamic. It is prudent to plan contingencies when interviewing elites (Ward and Jones, 1999). I could not get everything I needed from the website, as she suggested, and now I always prepare a key question in case I am only allowed to ask one quickly.

At the other end of the spectrum, one of my first encounters was a positive experience when a postgraduate researcher from Pune University (India) was present in one of my first interviews (occurrence 2 in Table 4.1). After great lengths to track down an apartment block that had been featured in a newspaper article due to its rainwater harvesting retro-fit, I turned up unannounced with two research assistants (for safety more than anything else). The other researcher was already there interviewing the society secretary. In this instance, it was useful as it meant that he had put aside time to be interviewed and to show her around the building and I was able to negotiate the situation for the benefit of my research, and I believe it was of mutual benefit for the other researcher. I had turned up uninvited and benefited from her arrangements. She was shy in asking questions to the distinguished respondent and I was able to ask questions relevant to both our interests. So, the positionality between researchers was important. I often wonder what that student thought of the encounter and whether it impacted her project. That first meeting made it possible to visit the same project several more times to establish a robust professional relationship and make more connections as well as see how other related projects were progressing. This demonstrates that having many researchers in a place can sometimes be beneficial for the researchers.

Positionality is considered by researchers thinking about the power dynamics between themselves and the people they are researching (with). Who the researcher is and their relationships with others will always impact the data gathered (England, 1994). I have shown in this section that those dynamics are often unexpected, extremely complicated, and ever shifting. The power, relationships, and positionality *between researchers* is also crucial to understanding over-researched places and experiences within them.

Ghosts of future researchers

The ghosts of future researchers are also present in how we collect data as responsible researchers and how we write about places. The (foreign) researcher is everywhere in Mumbai and has become a recognised part of everyday life in the city, which has led to questions concerning what effect this will have on my data, on the people I interviewed, on other researchers and projects, and on the city itself. There are academic researchers, documentary makers, journalists, and scouts for multinational companies all circling Mumbai and falling over each other to collect data. For the individual researcher, Mumbai is a rich enough source of information to last a lifetime/career, but we must consider the traces that we leave behind for future researchers. Mumbai is used as a case study in many research projects and shapes understanding of urban areas in the Global South, and particularly Asian cities.

The large body of literature on Mumbai is a useful resource for future researchers (for example, Pacione, 2006; Graham et al., 2013; Gandy, 2008) but it is easy to get bogged down in trying to consume and distil it all and attend all conference presentations on the city. This was demonstrated by an international workshop on urban infrastructures where three of the papers focused on Mumbai (and fourth Mumbai paper had been withdrawn) and it turned out that several people had interviewed the same respondents in Mumbai. The number of researchers does make data collection easier as it is an understood role and process (through direct access or media). If someone is interviewed regularly, then they understand the process and might be more open to it, as shown in some of the aforementioned examples. Our research can open up possibilities for future researchers but can also disrupt it. All data is biased and we must consider the complications of situated knowledge (Rose, 1997). Thus, reflexive thought is necessary to consider the researchers that will follow to create a more ethical approach. This means searching for and reading literature not just on our topics but going beyond our disciplines to look at the place, community, or people we research. This will make researchers more knowledgeable about potential over-researched places and give the opportunity to respond in advance. There are still barriers to the research process even in researcher-saturated cities. We need to carefully consider the negative (and positive) impacts on the people involved and adapt or mitigate accordingly, with participatory methods offering a way to build trust (Cochran et al., 2008; Cleary et al., 2016). Then think about the other researchers: what echoes they have left behind and what ghosts we leave for others. All this information should be used to influence where and how we conduct research and should be discussed openly. Too often, scholars only present methods as an aside or as a perfect process of moving from A to B without problems but revealing some detail of our research journeys might be more ethical (Horton, 2008). I call for more honest discussions of messy research processes to be normalised.

The Government of India is aware of the rising levels of researchers and the process of obtaining a research visa seems to get increasingly difficult every year. Research on tourism studies indicates that temporary urban residents (which could include many visiting researchers) affect the character and economics of a place (Colomb and Novy, 2016). Berlin and Barcelona both have protest movements mobilising against tourists and other visitors to these cities. This raises the questions of what it means to be researching a city that has an overwhelming concentration of visitors. A research visa must be registered within two weeks of arrival in India, putting serious pressure on the researcher to find somewhere to live (extremely difficult) as well as deal with yet more bureaucracy. The Government of India is attempting to regulate the research(er) situation and to increase the participation of Indian researchers by requiring affiliation with an Indian institute to obtain a research visa. Most, although not all, of the other researchers I encountered were also foreigners (from Europe or the USA) and these measures seems aimed at engaging with academic communities within Mumbai and India as a whole to take a more active role in researching the city. If affiliations encourage Indian institutions to take up more research and increase the impact of the research, and dissemination of the knowledge, then it is a positive initiative. Unfortunately, this may backfire by encouraging people to not apply for a research visa but to enter India on a tourist or business visa. The Government of India has subsequently reacted by requiring academics to supply a letter from their employer stating they are not undertaking any research or academic activities when applying for tourist visas. This shows the impact of past and present research practices on future researchers. I hope that it will encourage more collaboration between Indian scholars and international researchers in the future. This is a key way to create more appropriate and ethical research methods.

There has been consideration as to the impact of previous researchers, and to some extent contemporary ones, within the literature on methods and especially that on research fatigue. However, less consideration has been given to future researchers. We need to look at inter-temporal positionality. A reflexive approach is needed to consider how we will haunt future researchers so that we can fully account for this form of inter-researcher relationship in our practices.

We are all ghosts

Where we research and *how* we research has significant implications for the data collected and the subsequent theorisation. The everydayness of researchers and being-researched suggests the ghostly traces of researchers; expectations borne of past research experiences, encounters with other present researchers in interviews, and the management of future research through institutional responses to the prevalence of research. The number of researchers and projects targeting the same experts and communities for qualitative data collection has led to many respondents developing coping strategies. This could be by limiting impact on their time by pre-preparing answers to common questions and double-booking interviews or by capitalising on the interaction to create exposure for their own agenda. Urban theories are being generalised from a small subset of cities and, in the Global South, this is mainly megacities like Mumbai. This means that generalisations are being made from exceptional paradigmatic cities and there should be caution in applying these more broadly. This chapter has sought to demonstrate some of the effects that researcher ghosts have on researchers, data collection, experiences, and ideas.

The legacy of past researchers can lead to research fatigue and boredom, and thus we can also create this for future researchers. The literature on research fatigue consistently calls for reflexivity, and we again call on the research community to carefully consider the act of doing research. When undertaking research, we need to think about the power and positionality of ourselves as researchers. This is in relation to those we research (with) and about how we relate to other researchers, now and across different times. All these practices have a lasting effect on the places researched, on how researchers are reacted to, and also on what is written about in the literature. We must remember that where and how we conduct research not only shapes our own work but also the experiences of other researchers and their ideas. We must look at what has gone before, who is researching a place now, and the messages we leave for the future. We need to create new and varied patterns, including meaningful community engagement. The tale here is cautionary and researchers are urged to be reflexive in their future practices considering not only their own positionality but also the positionality of past, present, and future researchers.

References

Boesten, J., & Henry, M. (2018) Between fatigue and silence: The challenges of conducting research on sexual violence in conflict. *Social Politics: International Studies in Gender, State & Society, 25*(4), 568–588.

Button, C. (2014) *Domesticating infrastructure: Mumbai's middle-classes housing and rainwater harvesting*. Doctoral thesis, Durham University.

Button, C. (2017) The co-production of a constant water supply in Mumbai's middle-class apartments. *Urban Research & Practice, 10*(1), 102–119.

Clark, T. (2008) 'We're over-research here!' Exploring accounts of research fatigue within qualitative research engagements. *Sociology, 42*(5), 953–970.

Cleary, M., Siegfried, N., Escott, P., & Walter, G. (2016) Super research or super-researched? When enough is enough. . . *Issues in Mental Health Nursing, 37*(5), 380–382.

Cochran, P. A., Marshall, C. A., Garcia-Downing, C., Kendall, E., Cook, D., McCubbin, L., & Gover, R. M. S. (2008) Indigenous ways of knowing: Implications for participatory research and community. *American Journal of Public Health, 98*(1), 22–27.

Cochrane, A. (1998) Illusions of power: Interviewing local elites. *Environment and Planning A, 30*, 2121–2132.

Colomb, C., & Novy, J. (2016) Urban tourism and its discontents. 1–30. In Colomb, C., & Novy, J., eds., *Protest and resistance in the tourist city*. Routledge, London.

Crang, M. (2005) Qualitative methods: There is nothing outside the text? *Progress in Human Geography, 29*, 225–233.

Davis, M. (2004) Planet of the slums. *New Left Review, 26*, 5–34.

England, K. V. (1994) Getting personal: Reflexivity, positionality, and feminist research. *The Professional Geographer, 46*(1), 80–89.

Gandy, M. (2008) Landscapes of disaster: Water, modernity, and urban fragmentation in Mumbai. *Environment and Planning A, 40*, 108–130.

Gangar, A. (2003) Tinseltown: From studios to industry. In Patel, S., & Masselos, J., eds., *Bombay and Mumbai: The city in transition*. Oxford University Press, New Delhi.

Graham, S., Desai, R., & McFarlane, C. (2013) Water wars in Mumbai. *Public Culture, 25*, 115–141.

Hammett, D., & Sporton, D. (2012) Paying for interviews? Negotiating ethics, power and expectation. *Area, 44*, 496–502.

Head, E. (2009) The ethics and implications of paying participants in qualitative research. *International Journal of Social Research Methodology, 12*(4), 335–344.
Horton, J. (2008) A 'sense of failure'? Everydayness and research ethics. *Children's Geographies, 6*(4), 363–383.
Koen, J., Wassenaar, D., & Mamotte, N. (2017) The 'over-researched community': An ethics analysis of stakeholder views at two South African HIV prevention research sites. *Social Science & Medicine, 194*, 1–9.
Mandel, J. L. (2003) Negotiating expectations in the field: Gatekeepers, research fatigue and cultural biases. *Singapore Journal of Tropical Geography, 24*(2), 198–210.
McDowell, L. (1992) Doing gender: Feminism, feminists and research methods in human geography. *Transactions of the Institute of British Geographers*, 399–416.
Morris, Z. S. (2009) The truth about interviewing elites *Politics, 29*(3), 209–217.
Neal, S., Mohan, G., Cochrane, A., & Bennett, K. (2016) 'You can't move in Hackney without bumping into an anthropologist': Why certain places attract research attention. *Qualitative Research, 16*(5), 491–507.
Nijman, J. (2000) The paradigmatic city. *Annals of the Association of American Geographers, 90*, 135–145.
Omata, N. (2019) 'Over-researched' and 'under-researched' refugees. *Forced Migration Review*, (61), 15–18.
Pacione, M. (2006) City profile: Mumbai. *Cities*, 23, 229–238.
Pendse, S. (2003) Satya's Mumbai; Mumbai's Satya. In Patel, S., & Masselos, J., eds., *Bombay and Mumbai: The city in transition*. Oxford University Press, New Delhi.
Prakash, G. (2010) *Mumbai fables*. Princeton University Press, Woodstock.
Rao, K. (2011) Dhobi Ghat (Mumbai Diaries). Aamir Khan Productions, India.
Rao, V. (2006) Slum as theory: The South/Asian city and globalization. *International Journal of Urban and Regional Research, 30*, 225–232.
Rice, G. (2010) Reflections on interviewing elites. *Area, 41*(1), 70–75.
Robinson, J. (2006) *Ordinary cities: Between modernity and development* Routledge, Oxon.
Rose, G. (1997) Situating knowledges: Positionality, reflexivities and other tactics. *Progress in Human Geography, 21*(3), 305–320.
Roy, A. (2011) Slumdog cities: Rethinking subaltern urbanism. *International Journal of Urban and Regional Research, 35*, 223–238.
Singer, E., & Kulka, R. A. (2002) Paying respondents for survey participation. In National Research Council (2002) *Studies of welfare populations: Data collection and research issues*. Washington, DC: The National Academies Press. Ch. 4, 105–128.
Smith, K. E. (2006) Problematising power relations in 'elite' interviews. *Geoforum, 37*, 643–653.
Sukarieh, M., & Tannock, S. (2013) On the problem of over-researched communities: The case of the Shatila Palestinian Refugee Camp in Lebanon. *Sociology, 47*(3), 494–508.
Sullivan, M., Kone, A., Senturia, K. D., Chrisman, N. J., Ciske, S. J., & Krieger, J. W. (2001) Researcher and researched-community perspectives: Toward bridging the gap. *Health Education & Behavior, 28*(2), 130–149.
Taylor, S., Bills Walsh, K., Theodori, G. L., Jacquet, J., Kroepsch, A., & Haggerty, J. H. (2020) Addressing research fatigue in energy communities: New tools to prepare researchers for better community engagement. *Society & Natural Resources*, 1–6.
Ward, K. G., & Jones, M. (1999) Researching local elites: Reflexivity, "situatedness" and political-temporal contingency. *Geoforum, 30*(4), 301–312.
Welch, C., Marschan-Piekkari, R., Penttinen, H., & Tahvanainen, M. (2002) Corporate elites as informants in qualitative international business research. *International Business Review, 11*, 611–628.

5 La Duchère, Lyon, France

An over-researched place that ignores itself

Lise Serra

Introduction

From the perspective of the researcher, over-research modifies the way researchers may enter their field, the way they may interact with people within the area, and the way they should leave it. Knowing that a future research field could be an over-researched place might transform the way researchers will build their enquiries and research programmes. This is a hot point in urban research first because more and more people are leading enquiries: students, researchers, and journalists; secondly, because most of them will try to do it through the Internet on a wide scale. Anyone can carry out surveys from almost anywhere and have no idea of other people researching the same field at the same time. Being aware of acting in an over-researched place drives researchers to widen their research beforehand and take into account previous researchers as stakeholders as well as the traditional inhabitants or local authority representatives. These arguments are discussed with researchers, in the following: How do we build our own community to better understand urban dynamics? How do we invent new ways of field enquiries, closer to digital research? How do we adapt our methods to really big urban projects, without overlapping?

I tried to answer these questions when I came back to La Duchère, Lyon, France. By doing my PhD on La Duchère, a place undergoing urban renewal in Lyon, France, I suspected being part of something bigger than me, where more researchers were having a role. After ending my thesis, I explored this feeling and indeed listed 28 authors who had been writing from 1996 until today in 11 disciplines: from engineering to geography and history and from sociology and anthropology to urbanism and environmental sciences. I tried to gather them all and we finally met in April 2016 with six researchers. It gave birth to a new multidisciplinary research project on our non-wanted or unknown interactions on a common fieldwork where so many researchers have been working (Botea, Mongeard, and Serra, 2019).

I will here try to show firstly how La Duchère became an 'over-researched place', a place where researchers unknowingly interact and how they entered the field. We will see then how being an over-researched place can modify researchers' fieldwork and how researchers may create scientific bias in others' inquiry

DOI: 10.4324/9781003099291-6

process. Ultimately, I will look at how an over-researched place can be a place where new knowledge is created in confrontation between researchers and conclude with a proposal: how an 'over-researched' place can become a 'much-researched' place.

La Duchère represents a strategic position to reach the level of 'over-researched place'

La Duchère can be seen as an over-researched place from different points of view. First of all, it is near the centre of Lyon, a main university city where many researchers work and live, well connected to France and Europe by rail and roads. It is a research field easy to reach. Secondly, it is one of the first large French city projects launched by the Urban Renewal Agency (ANRU) in 2000, the first of a large series that has moved the French territory these past 20 years. Thirdly, the urban project team itself has called for research projects and always welcomed them. Operational stakeholders are easy to meet. This makes it popular to be included in lots of research as case studies (Barthe, de Blic, and Heurtin, 2013) in a pragmatist sociological point of view.

La Duchère is quite a large location, situated near the centre of Lyon, on a hill where fortified buildings used to protect the city from the west. A first neighbourhood was built at the end of the 1950s with 80% of social housing and 20% of private housing. Through the years, it has socially and economically weakened (Société d'Académie d'Architecture de Lyon, 2009). In 1984, the first building rehabilitation took place. In 1999, riots broke out following the death of a young inhabitant in a police action. Gerard Collomb was then the mayor of the ninth borough of Lyon, where La Duchère is located. As a rapid reaction, Raymond Barre, Mayor of Lyon at that time, looked for an urban proposal to renew the whole neighbourhood. Two years later, in 2001, Gérard Collomb was elected as the mayor of Lyon with, in his election proposals, the renewal of La Duchère. A special municipal team was in charge of the project, called the 'GPV team' for 'Grand Projet de Ville'. They are in charge of planning and following urban renewal operations and building demolition and constructions sites for the public contracting authority. They also create the link with every inhabitant who will have to move out before the demolition of their building. The team is plural, dealing with social, economic, and cultural project axes. As persons working there, they personalise the project itself.

The area is divided into four districts, covering 120 hectares with more than 1700 flats demolished and as many new ones rebuilt. In Figure 5.1 we can see the central place of La Duchère, with buildings to be demolished on the right of the picture, new buildings on the left and a tower to be renovated on the background. Demolitions concern social housing whereas new buildings present a mix of social and non-social housing with the aim of bringing the whole neighbourhood from 80% of social housing to 54%. The urban renewal project is sliced into two steps, beginning in 2000 and still not ended yet, 20 years later. In September 2013, it got the national sustainable label: 'écoquartier'. The construction site has expanded

Figure 5.1 April 2010, La Duchère, construction site.
Source: Photo by Lise Serra

rapidly with wide demolitions and plenty of families moved from one apartment to another and sometimes again to another.

Considering that the project is supported by Gerard Collomb, Mayor of Lyon from 2001 to 2020, with a lot of money invested, here are subjects for engineers, historians, geographers, urban planners, architects, sociologists, anthropologists, artists, economists, and so on.

The French Urban renewal national programme (PNRU) involves many public and private actors at the national and local levels: state services, city councils, social housing investors, and inhabitants. Their aim is to renew more than 500 neighbourhoods all over France for more than 4 million inhabitants with difficult living conditions. Demolition and reconstruction of residential buildings is the principal process in order to deeply modify the physical identity of such places. Mobility is the second axis to reconnect these neighbourhoods to traditional city centres and make it easier for inhabitants to go working or shopping. But this way of thinking about the renewal of post-war urban territories creates long and expensive projects. Projects face economic and social problems, technical and non-technical questions that can be analysed by researchers.

Counting 28 researchers working on this place from the end of 1996 until the present is not so unbelievable for such a big project but it makes this place a much-researched place. Thus, the range of works is wide: domestic social science (Halitim-Dubois, 1996), technical aspects of city renewal projects (Aubert and Toris, 1996; Brun and Casetou, 2014), conflictual representations within renewal housing projects (Botea, 2013, 2014; Rojon, 2014), demolition and inhabitants' movements (Overney 2014a, 2014b, 2019), ordinary citizenship (Deboulet and Lelevrier, 2014), construction sites and urban renewal (Serra, 2015), demolition process and urban renewal (Mongeard, 2018). These academic research studies are to be added to non-academic studies mainly linked to the urban renewal project. They both share sociological tools as quantitative or qualitative surveys, observations, and counting. In 2004, Catherine Foret, a sociologist harvesting inhabitants' memories for a research project, already spoke about the overexploitation of inhabitants' speeches by stakeholders. Indeed, with so many studies, some must have overlapped.

Many ways to enter the field

Since La Duchère is becoming a common context of many studies, we will see what that context can be. Here we pose 'context' as a complex set of relationships and a balance of individual and collective powers (Botea, Mongeard, and Serra, 2019). This context is different for everyone even if it seems to be the same place and time. We can think of the context itself as an experience, as a whole part of the enquiry work and not as an exterior data, something already here that we would just have to describe, from outside. Then, every additional study becomes part of the context, at least by the descriptions they present where it takes place. These descriptions are building a hypertext of La Duchère, adding to the context new points of view. Furthermore, researchers, when entering the fieldwork, are themselves becoming part of the field. Their choices, the way to lead the enquiry, are responses to the context. Research environment can here be illuminated by Tim Ingold (2011) and Augustin Berque (1990, 2000) works. Berque's oecumene and Ingold's environment place the researcher within a complex reality when everything is linked. Each researcher enters a fieldwork being part of it, transforming it as soon as they appear.

For example, starting points are all different and they imply different ways of leading research, even in the same location, with the same people. Indeed, every researcher has his own story of how they arrived in La Duchère:

1. BB, anthropologist, came to La Duchère through a research contract.
2. LM, geographer, did an exhaustive listing of every demolition workplace she could study. La Duchère was one of these and became one of her fieldworks for her thesis.
3. FG, geographer, worked in La Duchère as a civic servant within a student organisation and then made it as a fieldwork for her master's degree.
4. I, Lise Serra, urban planner, was looking for workplaces in Lyon. La Serl, one of the major stakeholders of the urban project of la Duchère and my thesis sponsor, proposed La Duchère with four other workplaces in Lyon. And so, it began to be part of my thesis fieldwork.

5 After that, a friend of mine, JC, architect and PhD student, was looking for a particular location to speak about architects and the horizon. I presented her La Duchère, gave her my contacts, and so it began to be part of her fieldwork.

In this sampling, we have five ways of coming to fieldwork:

1 through a contract, answering a call
2 after an exhaustive research of all similar locations within geographic limits
3 by being already there
4 by chance, this location meeting various items the scientific team needs to reach
5 through someone else

The second and the fourth ways could also form one item pointing the presence of relevant objects of study in the neighbourhood. The first and the fifth ways are already building the possibilities of an over-researched place, leading various researchers to a same place where people know that research is wanted and made possible.

An 'over-researched place', where researches overlap and may run into each other

With 28 researchers having worked on La Duchère, it is a well-researched place but can we say it is an 'over-researched' place? At which point could researchers have been stopped by others? On my fieldwork notebook, on July, 2015, I wrote:

> *The 2nd of July, year 2015, I remember having a coffee with LM on Place Abbé Pierre, which can be seen as the nerve centre of the new district, just after a building implosion. The place is crowded. We meet the 'GPV' team, people I was working with between 2010 and 2013, during my thesis intensive fieldwork. We join them and I present LM whom they haven't met. Shortly afterwards, the demolition team come for a coffee and see LM having coffee with the GPV team. They all are in a bad working relation the one with the others. LM stands up and leaves me but still, she will need some time to regain the confidence she had with the demolition team and make them sure she was not allied with the GPV team.*

Here, the status of 'over-researched' place is a place where two researchers being seen together may ruin their fieldwork. This is an important characteristic of over-researched places, totally absent from students' textbooks (Van Campenhoudt and Quivy, 2007; Beaud and Weber, 2010). Fieldwork techniques are often presented as if the researcher is alone in front of strangers, of non-researchers. Different techniques are presented to enter the field, to negotiate information, and to be part of the studied group. All these techniques will eventually fail if another researcher appears, reminding to everyone their own place in the social schema.

We can then ask if an over-researched place is a specific place for researchers? The aforementioned story shows how 'half-chance meetings' and 'half-planned meetings' are seen with different points of view. LM and I are both working on the same place. If we include demolition men and the GPV team, everybody at this café is here to work but with different goals and means.

On one hand, LM and myself are working on La Duchère as one case study within a bigger project called a thesis. On the other hand, the GPV team has worked here since 2001 and the demolition team is also working there for a long time before and after the decisive implosion.

For the GPV team, that used to work with researchers, we are part of the project, just as students, foreigners, city planners, and other visitors they welcome all year long. For the demolition team, that is not used to working with people from outside, LM had reached a state where she could almost be seen as one of them. After this meeting, it cost her to gain again their confidence.

This chance meeting highlights the strong bonds between most of the stakeholders on such long and big projects and how researchers may modify them. Here, by our friendship, LM and myself have slightly modified the relationship between the demolition team and the GPV team. As said by Beaud and Weber (2010, p. 248): 'The presence of the investigator acts as a catalyst on pre-existing social relations: It serves as revealing conflicts and local leaderboards. He becomes an issue in the balance of power that he does not know yet.'

It also shows how stakeholders are more or less connected to the outside, as we see in Figure 5.2.

This anecdote of unwanted consequences of our research meeting makes it possible to uncover relationships between actors that would otherwise have gone unnoticed. The tensions between contracting authorities and companies are known in professional circles but little studied by researchers. Researchers, through their presence and the multiplicity of roles they take on (Hayot, 2002), participate in the network of inter-knowledge within the urban project, making links between actors that were hitherto little visible. The analysis of these links enables a better understanding of certain aspects of the urban project, particularly its implementation process.

By bringing together doctoral students, we were breaking the distance that was set up between the project management and the project owner, which allowed them to continue working by limiting themselves to contractual clauses in a very tense general atmosphere.

As a very first indication, the long-time and wide perimeter of such urban renewal projects allows various researchers to work without overlapping. This means being aware of this research, even when they are in process, not yet published or not to ever be published (technical reports, students work, etc.).

Secondly, here the GPV team can be seen as key stakeholders for researchers: people who are easy to meet, having all the official documents, and holding the place history. The demolition team is less requested. Moreover, focusing on fewer requested stakeholders can be a line of inquiry into over-researched places.

Figure 5.2 Place Abbé Pierre, September 2011, the GPV general manager showing the place.

Source: Picture by Lise Serra

A second anecdote shows how the GPV team itself can stop some research directions.

In July 2010, as part of my thesis, I met the person in charge of communication in La Duchère. During the interview, she talks about an anthropology student who was due to deliver her study on the new inhabitants of La Duchère soon. For this person in charge, the study shows 'that the future inhabitants are not disappointed'. Indeed, by reading the report presented by the student, the communication manager understands that the interviewed inhabitants 'believed in the project, [that they] do not feel cheated by the promises that were made. [They] may have problems with developers on their housing delivery and the finishing touches, but [they] are quite satisfied with the whole project.' Yet, the final brief is less optimistic. The author highlights that the inhabitants are lost between successive moves and calls for participation in which they do not want to take part. They have doubts about their future: will they be able to stay, under what conditions? And the researcher concludes with a feeling of general distrust of public institutions. One year later, in September 2011, I spent a week in the GPV offices where

I regularly meet all the team's employees. Talking again about this anthropology work done earlier, my interlocutor's reaction is really violent: the student and her director have become, for my interlocutors, malevolent characters who make the inhabitants lie and who prevent the urban project from progressing properly. These reactions from people who are linked to my research project directly to the contracting authority make me stop my investigations on this anthropology research programme. This anthropology student and his director, by their work, became fully part of my fieldwork.

Looking back to field studies, a way to create new findings and new scientific problems

We could imagine that with so many researchers, we could build a long and plural history of a same project. From one research time to another, the field changes. Within the same space, people evolve. Institutions remain stable. Within institutions, people move from one position to another, from La Duchère to another renewal urban project. Inhabitants move outside or come inside. Every researcher meets mainly different persons but some of them would speak with everyone, like the GPV general director, the two architects-in-chief, some inhabitants who are always present in every meeting.

On the one hand, when researchers meet, they exchange on history, each one adding sources, data, or anecdote. But at a higher level, it seems harder to build a common unique history. This would not be a natural way for every researcher involved at least because researchers are not a compact, one-way heading group. Furthermore, there is, nowadays, no need for such a history and there are already so many tales made by as many stakeholders. But parts of history are something researchers have in common.

What else is common between most of us, researchers? A common observation of every researcher working on La Duchère is that it is a hard place to work in. It is first marked by politics at local and national levels; secondly, it is marked by social difficulties as poverty and illiteracy; and thirdly, it is marked by segregation and urban isolation. These difficulties are challenging researchers' 'normal' tools that they will have to reinvent, particularly challenging the very place of the researchers within the field. First, at the local authority level, some central persons are quite hard to deal with, for various reasons. That is something researchers can talk about and give them clues about how to deal with them in the best way. Secondly, linked to poverty, illiteracy is also linked with the lack of knowledge in reading graphical representation like reading a masterplan, for example. Urban representation often uses master plans to show how the construction site modifies the bus path and where the shopping centre is temporarily moved. However, most of the people living there could not read a plan. Such graphic representations must be reinvented if inhabitants' points of view are wanted for the study. Thirdly, segregation and bad mobility amenities make it more difficult to go there often and at every hour of day and night. Then, if various researchers are working at the same time, they can even share a car ride or exchange tips on special times and dates when it is needed to be there.

A third finding is the hybrid nature of our scientific productions. This can be seen as a result of creating a new way of understanding urban studies and their tools and processes but also be seen as a new scientific problem. Between observations and factual data, concepts, tales, experience stories, photos, shared surveys, and the mix of these tools create a scientific process that is hard to make clear as a perfect theoretical material. It has to be accepted that this science is always in the process, not finished, and deeply linked to authors, times, and locations. Fieldwork close notes, photos, and videos are widely used to share people's words, interactions, sensations. These survey materials become part of the knowledge created about the transformations in the neighbourhood.

Last but not least, La Duchère seems to have such a strong aura that fieldworks are becoming synonymous with urban renewal projects. Speaking of 'working in La Duchère' for a researcher bounds him to the place and all the stereotypes of it. In a scientific conference, you will be represented as someone 'working on La Duchère'. Occasionally, someone will call you to organise a guided visit of the district for friends or colleagues. Even if the fieldwork is not as wide as the urban project perimeter and even if it represents a small part of the whole research project, it colours the research and the researcher with all the surrounding themes of a well-known place. That is also a direct consequence of an over-researched project.

Conclusion: manifold researchers

Here, the over-researched place can be seen as a positive point, where researchers can confront their methods and results. It can also be seen as a negative point because precedent researchers may have made much more difficult access to some part of the field by their work, publications, and presentations. Researchers, by their number, are being too visible when often looking for invisibility. They fully become a part of the fieldwork and must take it into account.

Researchers, being nearby stakeholders, become stakeholders themselves and vice versa: stakeholders being accustomed to researchers are modifying their own points of view. The urban renewal field study becomes a place of resources, links, actions, and cognitive universe categories, where the researcher is only one part of the knowledge construction. This hypothesis also deals with an intimacy dimension well shown by Jane Favret-Saada (2009) to the research process that is hard to analyse. Over-researched places permit confronting different researcher stories in order to go further in the comprehension of this intimate dimension.

One way to conclude would be to list some lessons I learned from the La Duchère experience:

- Beforehand, check if the field might be already over-researched or could become one in the near future;
- If so, do not stop yourself but prepare a strategy to team with people already there;
- Use this field characteristic to think of new ways of enquiries;
- Think of your work as part of the project you study, including as a political engagement if needed.

As the first postdoctoral researcher here, going back to the field with other researchers showed me how the thesis can be a unique and fundamental work, building ourselves as researchers and also how every research modifies us and our point of view. A 'much researched' place permits confronting works and ideas to multiply the points of view and expanding experiments and knowledge about a location but furthermore about ourselves. At the end of my thesis, I tried to show how cities are in constant modification and how this should be part of their definition. After this piece of work, I should take into account that researchers are also in constant transformation and that many researched places are good places to observe it.

References

Aubert, Christian and Toris, Jean-Luc (1996) "Le boulevard périphérique nord de Lyon: GIE Lyon Nord – projet ouest: 2, la tranchée de Vaise", *Travaux*, 1996–9, n°773, Marne-la-Vallée, Fédération Nationale des Travaux Publiques & des Synd Aff, pp. 39–42.

Barthe, Y., de Blic, D., Heurtin, J., Lagneau, É., Lemieux, C., Linhardt, D. and Trom, D. (2013) Sociologie pragmatique: mode d'emploi. *Politix*.

Beaud, Stéphane and Weber, Florence (2010) *Guide de l'enquête de terrain*. Paris: La Découverte.

Berque, Augustin (1990) *Médiance, de milieux en paysages*. Paris: Belin, coll. "Reclus géographiques".

Berque, Augustin (2000) *Écoumène. Introduction à l'étude des milieux humains*. Paris: Belin.

Botea, Bianca (2013) "De quelques usages de la valeur dans les recherches anthropologiques: un regard sur le développement urbain (Jimbolia, Roumanie)" in Baillé, Jacques (dir.), *Du mot au concept. Valeur*. Grenoble: PUG, pp. 85–101.

Botea, Bianca (2014) "Expérience du changement et attachements. Réaménagement urbain dans un quartier lyonnais (La Duchère)", *Ethnologie française*, n°3, vol. 44, pp. 461–467.

Botea, Bianca, Mongeard, Laëtitia and Serra, Lise (2019) "Connaissances par proximité dans la recherche sur la rénovation urbaine", *EspacesTemps.net*, Traverses, 8 November 2019. www.espacestemps.net/articles/connaissances-par-proximite-dans-la-recherche-sur- la-renovation-urbaine/

Brun, Alexandre and Casetou, Evariste (2014) "Renaturer les rivières urbaines. Le projet du ruisseau des Planches à Lyon", *Métropolitiques*, 8 janvier 2014.

Deboulet, Agnès and Lelevrier, Christine (dir.) (2014) *Rénovations urbaines en France*. Rennes: PUR, 360 p.

Favret-Saada, Jeanne (2009) *Désorceler*. Paris: Éditions de l'Olivier, coll. "Penser/Rêver".

Foret, Catherine (2004) "Les productions mémorielles dans l'agglomération lyonnaise", in Martine De Boisdeffre et Claude Brévan (dirs.), *Colloque "Villes et mémoires"*. Paris: Éditions de la DIV, pp. 62–65.

Halitim-Dubois, Nadine (1996) *La vie des objets: décor domestique et vie quotidienne dans des familles populaires d'un quartier de Lyon, la Duchère, 1986–1993*. Paris, Montréal: l'Harmattan, collection Logiques sociales, 301 p.

Hayot, Alain (2002) "Pour une anthropologie de la ville et dans la ville: questions de méthodes", *Revue européenne des migrations internationales* [En ligne], vol. 18—n°3 | 2002, mis en ligne le 09 juin 2006, consulté le 16 novembre 2018. http://journals.openedition.org/remi/2646

Ingold, Tim (2011) *Being alive: Essays on Movement, Knowledge and Description*. London: Routledge.

Mongeard, Lætitia (2018) *De la déconstruction au recyclage des déchets: analyse socio-économique et territorialisée de la filière démolition dans l'agglomération lyonnaise*, thèse de doctorat en géographie de l'École Normale Supérieure de Lyon, sous la direction de Vincent Veschambre.

Overney, Laetitia (2014a) "Par-delà 'la participation des habitants': pour une ethnographie de la petite politique. Le cas d'un collectif d'habitants de la Duchère", in M. Carrel and C. Neveu (dir.), *Citoyennetés ordinaires: pour une approche renouvelée des pratiques citoyennes*. Paris: Khartala, pp. 131–166.

Overney, Laetitia (2014b) "L'épreuve des démolitions à la Duchère: tactiques de résistance d'un collectif d'habitants", in Deboulet, Agnès et Lelévrier, Christine (dir.), *Rénovation urbaine en Europe: quelles pratiques ? Quels effets ?*, Presses Universitaires de Rennes, collection Villes et Territoires, pp. 125–134.

Overney, Laetitia (2019) "Agrandir la parole des habitants", *EspacesTemps.net*, Traverses, 4 October 2019. https://www.espacestemps.net/articles/agrandir-la-parole-des-habitants/

Rojon, Sarah (2014) "La rénovation de l'habiter dans le grand ensemble de la Duchère. Pour en finir avec la figure des 'nouveaux habitants'", *Recherches sociologiques et anthropologiques*.

Serra, Lise (2015) *Le chantier comme projet urbain*, thèse de doctorat en urbanisme et aménagement de l'université Paris Ouest Nanterre la Défense, sous la direction d'Hélène Hatzfeld, 500 p.

Société d'Académie d'Architecture de Lyon (2009) *La Duchère, 1948–2014*. Lyon: Bulletin de la Société d'Académie d'Architecture de Lyon, n°17–18, octobre 2009, 75 p.

Van Campenhoudt, Luc and Quivy, Raymond (2007) *Manuel de recherche en sciences sociales*. Paris: Dunod.

6 'Research has killed the Israeli-Palestinian conflict'

Navigating the over-researched field of the West Bank

Alejandra de Bárcena Myrsep

Introduction: the West Bank as an over-researched field

Since its outset, research has been documenting, analysing, and disseminating the events of the Palestinian–Israeli conflict. Yet, the conflict still pursues, consolidating the West Bank as a paradigmatic field. Consequently, working on this conflict is frequently referred to, within research circles, as a career catalyst: positive for the researcher's own career but negative for those involved. In effect, the accessibility, funding opportunities, and constant interest in the conflict have typified the West Bank as an over-researched field. In reflexively analysing my enabling position in the over-researched issue—a foreign researcher conducting short-term research on Bedouin communities—I intend to examine a troubling question through autoethnography: how can research kill a conflict? How has research become a colonial continuity in the West Bank? To this end, two concepts relating to over-research, the *burden of occupation* and the *futility of research*, will be examined through a critical lens.

In this chapter, over-research is conceptualised within an ethical framework in relation to agency, profiteering, and vulnerability. By using *over-research* as a conceptual label—and admitting its multifaceted and often contradictory and problematic nature—this chapter analyses the emergence of such claims as a by-product of the NGOisation of the West Bank's popular struggles and social movements. NGOisation, as a theoretical framework, critically addresses the demobilising effect of the institutionalisation and professionalisation of NGOs (Choudry and Kapoor 2013). In the context of the West Bank, NGOisation is applied as an umbrella term for NGOs and non-profit organisations (NPOs). This is because, though different entities, NGOs and NPOs fall under the same registration requisites of the Palestinian Ministry of Interior and must adhere to the Palestinian NGO code of conduct which requires amongst others transparency, good governance, and accountability. These Palestinian non-state and private actors are addressed in this chapter as PNGOs. Non-governmental organisations (NGOs), propelled by the neoliberalist agenda, must compete against each other for funding and resources, essentially, making them contingent to vertical clientelism to satisfy donors' agenda, budget, and delivery expectations (Hajjar 2001).

DOI: 10.4324/9781003099291-7

Accordingly, an over-research analysis becomes an ethical probing of the motivations and methods behind the researcher's interactions (Koen et al. 2017).

The term 'over-research' initially emerged in Urban Whitaker's essay titled: 'The Dangers of Over-research' where he described over-research as an *'overkill emphasis'* embedded in 'the fact that we already know enough.' (Whitaker 1963, 68). His argument was based on the belief that *enough* had been researched to offer peace solutions to the Cold War nuclear crisis and so the next step for researchers should be to step back and find solutions. He elaborated that 'over-research' is the consequence of the academic greed for career advancement, where academics find prestigious and financial success in research-orientated work and publishing. It rendered teaching as the lower strata of the academic hierarchy (*ibid.*). Insofar that Whitaker stated that 'we are paying most for what we need least [and] honor least those whose efforts we need most' (*ibid.*, p. 69). It makes a crucial point in my argument that over-research is a consequence of the NGOisation of the West Bank, where international donors are paying for what we need least: a system based on meritocracy, grantsmanship, and 'officialdom'-orientated research (Jellinek 2003). Inasmuch as NGOisation has enabled a system of over-research by eager, over-zealous Palestinian NGOs (PNGOs) fail to act as gatekeepers to protect the interests of the Palestinian research participants and project beneficiaries. I will delineate how international organisations (IOs) and NGOs take credit from Bedouin community-built buildings and success stories thus straining the relationship between those researched and the stakeholders surrounding the knowledge-production apparatus.

At its core, over-research emerges due to a conflict of interest between stakeholders and their expectations *for whom* research is useful and relevant. Within this enabling system, over-research illustrates an overabundance of NGOs and researchers conducting participatory-based projects in specific and 'closed contexts' (Koch 2013), largely short-term and managed by a top–down structure. In the case of the West Bank, an *(inter)national elite*, largely *urban middle-class* is researching *marginalised, low income*, and *vulnerable* communities. Then, over-research becomes a form of on-site data saturation, that is to say, that no more beneficial research is achieved. Therefore, research projects are forced upon local communities to the extent that the plethora of research projects are associated with the notions of invasive methods that perpetrate more damage than good to the researched groups.

The realisation behind the over-research and NGOisation analysis lies within my Palestinian academic mentor's contradictory stance towards research: hopeful yet pragmatic; passionate yet apathetic; and goal-driven yet defeated. There were instances of shared honesty where we would question the usefulness of research to deter the occupation and conjecture that the West Bank and Gaza Strip are an experimental Petri dish of settler colonialism, positioning Palestinians as the 'natural objects' of conflict research. Perhaps, it is then appropriate to express that the fetishism of research is in the search for authenticity; yet, there is a lacking intervention within constant intrusions. Thus, the researcher probes the emotional aspects of conflict, in an attempt to reach an inherent truth about humanity.

However, the researcher's authenticity has been corrupted by own intrusion, dried up long ago by countless researchers who came before with the same questions, entitlement, and questionable motivations.

In a conflict termed as 'intractable', somewhere amongst vast forms of policy writing, resolutions, conferences, focus groups, and immeasurable data collection, research became an accessory to the occupation's dynamics. I will argue then that research has become a futile endeavour, positioning the burden of the occupation on those required to participate in the knowledge production apparatus. Subsequently, the circus of the Palestinian struggle for self-determination continues to attract both foreign researchers, voyeurs, and tourists. In its path, the promotion of participatory-based methods solidifies the perception of being over-researched among Palestinian communities, effectively rendering researched communities as commodities rather than active participants. This became evident as a young Bedouin activist who exclaimed:

> And you know, *'I'm sorry and I am in solidarity with you,'* this is not enough anymore. Okay you are in solidarity with me, *thank you*. You are sorry, okay, but what else? The people have to act for justice. You know, some researchers come here just to live like adventure . . . experience!
>
> (Interview with Young Bedouin representative, July 27, 2019)

Therefore, the over-researched Bedouin communities in the West Bank, who are constantly targeted as an issue-orientated group, perceive research as the justification for adventure rather than to genuinely aid their struggle. Nevertheless, there is an inherent contradiction within the West Bank's over-research issue and that is the benefit of research to attract advocacy and funding, leading to over-research. As the young Bedouin explained:

> *Wallahi*, I will be honest with you and advise you. Don't read the books. Why? Just listen to the people who live their experience. You will hear from them many things that are not in the books.
>
> (Interview with Young Bedouin representative, July 27, 2019)

This palpable incongruity demonstrated the West Bank's paradigmatic nature. Consequently, this chapter wishes to address an under-analysed aspect of the over-research issue: the research fatigue of Palestinian scholars, or more explicitly, their burn-out crisis concerning the lack of decolonizing progress in institutionalised research. As my mentor explained point-blank on my first day: '*research has killed the Palestinian-Israeli conflict*' (personal communication with Palestinian mentor, July 18, 2019). As I would struggle to schedule interviews and focus groups, my mentor gave me another crucial piece of the puzzle: '*everything has already been written. You can read what has already been researched and rewrite it to make it your own*' (personal communication with Palestinian mentor, July 18, 2019). Yet, borrowing Gita Verma's analogy (2002), the golden goose keeps on giving its golden eggs in this industrial supply of artificial knowledge, stimulated by researcher's 'fetishism for authenticity'. And so, as Verma illustrated, 'no one wants to kill [the golden goose]' (Verma 2002, 34), encouraging the pilgrimage of more researchers to take more.

My hope is that my fieldwork experience will inform other students' choices when considering the West Bank as a research field. This aim fits into the efforts of increasing the transparency surrounding fieldwork in conflict regions (Browne and Moffett 2014). Most importantly, however, it is my wish to address the ethics surrounding the West Bank's politics of research, from research scams to unethical practices and dubious advice, which I believe mirror the over-research issue against a backdrop of NGOisation.

Disposition

To demonstrate the premise that the West Bank is an over-researched field, this essay will first address the constraints posed by this methodological challenge, shaped by gatekeepers, research fatigue, and research resistance. It will then delve into a short description of the West Bank followed up by a focus of the Bedouins case. Next, I investigate the effects of the Oslo Accords and the subsequent embrace of the neoliberalist agenda on the grassroots activism in the oPts (occupied Palestinian territories). The latter fosters a dialogical conversation in the role of NGOs as gatekeepers in the over-research issue as promoters of research as a lucrative business model. In this discussion, I will address who directly benefit from the research projects and the NGO industry.

Overall, I intend to showcase the differing perspectives of my informants to present and justify my argument. Accordingly, I will rely on cultural analysis to unravel the patterns of behaviour with research and NGOisation, which will be supported by reflexive auto-ethnographic accounts from the field.

Methodology

This research is based on three-months of fieldwork in the West Bank from July to October 2019, while conducting ethnographic research for a Palestinian non-profit. I have consciously kept the research institute and its location anonymous due to a confidentiality agreement.

Arriving as a student volunteer researcher, I was commissioned to produce a comprehensive report that would fit in with the non-profit's interests. Their work had a focus on a multitude of development programmes which branched into specific departments. Their projects were predominantly funded by IOs, donor countries, and the Palestinian Authority (PA). Consequently, they centred around policy writing, reports, programmes, conferences, and education and development projects. The Bedouin ethnic displacement was chosen believing that the research institute would facilitate the accessibility to the field considering that they had ongoing rural development projects. As a first timer in the West Bank, I did not have a network, no means of transportation, and no proficiency in Arabic. As a gateway, my colleagues suggested that I send emails to other NGOs. After spending three days, only two replied. From those, one was a research scam. My unanswered emails are not an isolated case (see Browne and Moffett 2014) as researchers constantly bombard possible gatekeepers to access the field.

As a result of these logistical constraints, the qualitative data sample was small and seemingly poor due to various forms of research resistance where emails were ignored and interviews were cancelled, and oftentimes, a successful scheduled interview would develop into a fruitless interview. By fruitless, I mean rehearsed answers, no follow-up interview, and nuances of speech lost in translation.

Nonetheless, the primary source of material for my analysis is based on *participant observation* while working at the Palestinian research institute during seven *semi-structured interviews* with Palestinians, evenly spread between urban and rural, as well as refugee camps, *back-room note-taking* at two focus groups, and two NGO *go-alongs* to Bedouin communities. Three interviews extended to other gatekeepers which included one IO, an INGO, and one activist working for Bedouin rights.

Participant observation and go-along (Kusenbach 2003) opened the opportunity to delve into the inner intricacies of the politics of research. Over-research became apparent due to my outsider-insider status at the research institute, where: 1) I had a different set of stakes in the research enterprise; 2) I was self-funded, which meant that my research was not grant-driven; 3) I was inexperienced conducting fieldwork located in conflict zones; and 4) I was not an active part of the research institute's on-going projects, but I was encouraged to be an observer. This insider experience led me into an interpretative path. If I had just interviewed the research NGO, I would have received a curated frontstage. However, while volunteering as a student researcher, I could interpret the backstage and make sense of using insider knowledge. Hence, I would have been unable to grasp the over-research phenomenon if I had not been a participant-observer of the practices surrounding research in the West Bank. Accordingly, the implications of an over-research field first became evident as I was given a cubicle and a computer to conduct armchair research. Given the situation, I had time to consider the underlying reasons behind my struggles to access the field and the apparent failure to 'break ground' with my informants. The following fieldwork diary entry encompasses my experience as an early-career researcher in the West Bank: *'when to give up on a field that rejects you'* (author's fieldwork diary, September 10, 2019).

Fieldwork diaries and informal conversations shaped most of my analysis, as they retrospectively showed my difficulties in the field and the discourse behind NGOs' work: their performed *frontstage* and *backstage* (Goffman 1956). As Browne states, diary writing has a catharsis while conducting fieldwork in Palestine when considering the researcher's perceived inertia and powerlessness to the conflict's resolution (Browne 2018). Keeping a diary, in contrast to research-focused fieldnotes, helps maintain these emotional aspects of research (Browne 2018) which became apparent when I continually reference my growing uneasiness towards positionality and the unbalanced power dynamics engaging with a vulnerable community.

As such, the use of auto-ethnography as a methodology is a strategic processing of the subjective experience (Adams et al. 2015). As experiences are inherently personal, the declaration that the West Bank is an over-research field can be only classified as an interpretation of the field. By reflexively analysing my enabling position, I am exposing my own participation in the over-research issue. This is why auto-ethnography is a central part of my methodology, as reflexivity examines the convoluted and confusing character of fieldwork and its unexpected results (Browne 2018). Thereby, allowing the articulation of our subjective interpretation.

The overall challenging fieldwork in the West Bank informs the analysed issue of over-research. If the fieldwork had been trouble-free and undemanding, I would have most likely focused on adding to the extensive literature of Bedouin displacement, rather than underlying the issue of Bedouin over-/under-research. Sukarieh and Tannock (2012), who recognised the over-research issue in the closed context of Palestinian refugee camps, declared that perhaps the solution to over-researching would be to admit that no more research is needed. In practice, research designs should account for the potential harm to over-researched communities.

Background

The partition of Palestine was prompted by Resolution 181, in which the newly established United Nations (UN) advised a two-state solution with Jerusalem as a neutral international city. The West Bank's territorial entity steadily came into being after the Israeli Declaration of Independence, prompting the 1949 Palestinian Exodus (*Nakbah*) and its annexation to Jordan's sovereignty, demarcated by the 1949 armistice line (the Green Line). As a result, 750,000 Palestinians concentrated in the Gaza Strip, the West Bank, and the neighbouring Arab countries as a result of their forced expulsion and displacement of the war (Pappé 2006).

Palestinians have continually struggled to control their narrative as a people and recognition for statehood. Therefore, Palestinians grapple to maintain a monopoly of knowledge as they continue to compete against Israel's lobbying, Zionist ideo-religious 'truth claims' and economic and military superiority (see Khalidi 1997). National identities affirm both their belonging and historical continuity on the land, which brand the conflict as *intractable* and its past as *irreconcilable* (McDowell and Braniff 2014).

The oPts (including East Jerusalem) are recognised as occupied territories under international law; however, Israel contends it by labelling a 'disputed territory' with the legal definition that there has never been a formal state of Palestine since the territory was first under Ottoman rule and later under the British Mandate (1920–1948). In 1988, the Palestinian Liberation Organisation (PLO), a union of several political parties and popular movements, declared officially the independent state of Palestine.

Case study: the 'Bedouin question'

The Arab Bedouins of the West Bank are considered a Palestinian sub-grouping in the development aid discourse which entails their targeting as an issue-orientated group. This minority subgrouping is up for controversy as identity classification separates Bedouins from mainstream Palestinian discourse and aims (Amara and Nasasra 2015), in effect, isolating them into a cultural margin. As I will demonstrate further along, this attitude is often fostered by NGO marketing and projectisation strategies.

This systematic targeting positions Bedouin at the forefront of the occupation as Bedouins fall in Area C category. In effect, Bedouins sustain high levels of excessive research that according to their accounts varies from time to time. This depends according to their level of immediate threat. The Bedouin community undergoes relative periods of isolation until the threat of demolition is unavoidable and so activists, scholars, NGOs, and media, overcrowd the village to take photographic and video evidence. Verma criticizes NGOs' involvement in Indian slum evictions because of the 'glamorous' status they provide as evictions rally the most support and attention. As she expresses: 'It may stop the occasional bulldozer but, for the rest, it does little beyond change the label from "problem" to "solution" with some creative jargon in small print' (Verma 2002, 74). In the meantime, the threat of demolition remains constant, a Bedouin informant stated while pointing to the opposite hill, where a hundred Israeli settler units are meant to be built. As residents of Area C, Bedouins too are denied building permits and basic infrastructure, and therefore, do not have direct access to water and electricity supplies.

Bedouins engage in high levels of research as means for advocacy and recognition to the extent that research becomes a necessity in order to gather assistance and support against the Israeli occupation. This becomes crucial when considering their lack of recognition amongst urban Palestinians: 'Unfortunately, the internationals who know more about my village are more than the Palestinians' (interview with young Bedouin man, July 27, 2019).

While research may be considered an act for justice and provides incentives such as: 1) assistance; 2) cathartic effects; 3) economic gain; and 4) activism (Clark 2008, 401), it often emulates forms of exploitation. Crucially, it entails a process of trauma revival in which Palestinians engage in constant probing and questioning about the traumatic aspects of the occupation. This adds to the high levels of emotional association with research and the burden of the occupation. This leads to an association of over-research and intrusion, more visible in the connection of research as a colonial continuity of debilitating foreign encroachment. Therefore, for Bedouin's *sumud* (steadfastness) on the land, the burden of occupation becomes a paradigm (Khun 1970) between an overriding necessity on research for advocacy, notwithstanding its benefits and implied imbalanced power dynamics, and its admission that research will do little to improve their precarious situation.

Over-research as a result of NGOisation of the West Bank

It is not a secret nor a novel realisation that the Palestinians are over-researched. Indubitably, the West Bank holds the status of 'symbolic location' (Neal et al. 2016, 497) due to its religious, political, and historical significance, which entices and lures countless researchers. Edward Said stated that the Israeli–Palestinian conflict has been an object of obsession and fascination within Middle East 'specialists' since the Second World War (Said 1993), while Browne & Moffett described over-research as an alluring *'fetishism of conflict'* which results in the Palestinian communities' 'frustration with western researchers' apparent fetish for conducting research . . . in war-torn regions that ultimately generates few tangible results on the ground' (Browne and Moffett 2014, 230–231).

While researching in Bedouin communities in the West Bank, I would often encounter an assortment of researchers from social sciences to engineers and doctors, overlapping and interviewing the same vulnerable communities. Most of them would represent a prestigious organisation. The village then would transform into a frenzy of activity, a messy arrangement of questions and answers in English and Arabic, while children were auscultated and drawn blood samples, a hygiene workshop was taking place behind us, and my organisation discussed maps with a community representative. The over-research, evident in the swarming activity, took over every day in the village.

Albeit constricted by the occupation, Palestinian civic organisations have proliferated since 2000 with their registration increasing each year in number (MAS 2007, 12). This occurred due to the precipitated increase of external financing after the Oslo Accords. According to De Voir and Tartir (2009), the oPts have received an increase of 600% external aid funding between 1999 and 2008. By 1999, the total of US$48 million increased to US$257 million by 2008. The flow of aid led to the uncontrollable burgeoning of NGOs in a concentrated area. In 2007, MAS estimated that the number of NGOs amounted to 1495 in the oPts, a 61.5% increase since 2000 which results in a high ratio of NGO per citizen, a total of 2848 citizens per organisation (MAS 2007). The high amount of NGOs bear evidence of the root cause of over-research in the West Bank as its effects are visible in the proliferation of professionalised Palestinian NGOs (PNGOs). Gita Verma (2002, 74) described it as:

> The changing role of NGOs—from activism to convergence with government, from identifying problems to becoming partners in state-designed solutions, from giving voice to the people to becoming a hand for universalizing global paradigms, from minimal operations to massive fund-raising, from simple folk to celebrity.

To provide a background, PNGOs' authority lies in their historical and active role as grassroots organisations defending Palestinian civil rights since the British Mandate. In absentia of a functioning central government pre-Oslo Accords, voluntary grassroots organisations filled in the role of service providers, representing a united Palestinian national struggle against the occupation and displaying high levels of inclusion, reliability, and capability (Hajjar 2001). The establishment of the PA led to local tensions and the displacement of previous responsibilities withheld by grassroot groups, which sparked an environment of competition over funding and spheres of influence (MAS 2007). This increased as the PA attempted to settle dominance over state and civil society. Consequently, rivalry and disunity arose. Grassroots organisations sought out foreign donors as means for their survival which lead to their enforced professionalisation (Hajjar 2001). At its core, the PA argued that grassroots organisations threatened national unity (*ibid.*). However, I will argue that it is the PA's embrace of neoliberalist intervention on domestic markets that opened the door for foreign encroachment. Which is illustrative in the PA's corruption allegations and repressive response against its critics. This demonstrates that over-research, as a form of research fatigue, arises from these critical changes in civil society.

Though PNGOs may attempt to maintain 'good faith' research, the reality is that through professionalisation, there is a withdrawal from the field and a detachment from a participatory struggle. Lea Jellinek describes this process in the context of Jakarta as an empowerment grassroots project transformed into a 'complex, top-down, technically orientated, capital-intensive bureaucracy guided by government and big international donor agencies' (Jellinek 2003, 179) which was essentially 'out of touch with what was happening on the ground' (*ibid.*). A feeling I often associated with my research institute as I heard other NGOs talk about their disengagement attitudes towards their projects.

The expansion of an NGO leads to official recognition, which in turn attracts funding, prestige, and power. However, Jennick described it as the paradigm of the neoliberalist success: once a grassroots organisation attains a certain level of prestige and status, a conformist attitude falls in place. As a result, the NGOs push aside their grassroots orientation and replace it with official ceremonies, awards, and shiny plaques (*ibid.*). The lack of exchange of ideas makes 'expertise' claims deceitful, enabling a 'consumptive' environment where resources and time is wasted in patting the powerful on their backs (*ibid.*). It sets off a scene of unequal trade-offs between NGOs and project beneficiaries which leads to their fatigue and refusal to engage in research, deeming their engagement futile to improve their current struggles.

As donor agendas are the guiding torch, donors may have demanding and unrealistic expectations, pushing NGOs to cater to the stakeholder's preconceived expectations, which oftentimes are based on disassociated notions of the field. As a result, popular topics are prioritised to attract donors, while other issues do not make it into the literature or qualify to receive services, thus becoming 'under-researched.' This bears the argument that under/over-research follows the dynamics of this donor funding, and can be seen as popularity contests in donor's agendas.

I will argue that over-research claims are based on the evidence that certain IOs are taking credit from community-built projects that were built in the 1950s without granting recognition to the community (personal communication with Bedouin activist, August 16, 2019). This credit-claim strategy is based on using a kernel of truth to manipulate the facts, which fosters an environment of foreign outsiders piggybacking off local merits and rendering an image abroad of agentless Palestinians. This fosters a system of meritocracy and 'officialdom' whereas PNGOs' main objective is to maintain their reputation and grant income.

Grantsmanship (Deloria 1969) is the practice of marketing and proposing projects to secure grants, irrespective of its actual necessity. The project does not necessarily have to be beneficial. The project's deliverables can be marketed as '*success stories*' through rhetoric. This is because donors hold the reins as projects are threatened by potential failed evaluations that can lead to their discontinuity (Cooley and Ron 2002). This uncertainty unfolds into an enabling system of data falsification, where project reports camouflage failures to protect the NGO's interests, especially when their organisational viability is at stake (*ibid.*). Consequently, communities are considered commodities to compete for and transfigured into roles of passive beneficiaries instead of active participants (Arda and Banerjee 2019). Therefore, projects in the West Bank tend to misrepresent the beneficiaries and their interests, where PNGOs and INGOs justify their projects through invention or exaggeration in order to shoehorn applicable solution-based approaches, and consequently, market their projects (Ferguson 1990).

To achieve goals and quotas to appease supervisors and donors, fieldworkers may be compelled to engage in ethical malpractice. I encountered this when an acquaintance from a development INGO admitted they interview children over adults, as adults tended to exaggerate accounts for the sake of encouraging more benefits from the INGO. Children in comparison were more inclined to tell the truth (personal communication with an INGO researcher, August 3, 2019). This unethical practice by the INGO to which my acquaintance admitted finding trouble is not unique to the West Bank. It seems to permeate into global research practices. As Khan records: 'I have sat on research meetings where nothing unethical could be felt about attaching recording microphones to children in *madrasahs* [schools], so that what they talk about could be researched' (Khan 2015, 108). These academic attitudes display an entitlement, as subjecting children to the research process is based on the unbalanced dynamics between the adult researcher and child participants. By interviewing dependents, the research ethics are compromised in the name of 'truth' or quality of data. It is ethically dubious because it denies valid and informed consent. It can even jeopardise the wellbeing of the children as they are unaware of the implications of their participation and 'telling the truth' can cause conflict with the adults. This early experience with research can have long-lasting negative associations, which culminates in the forms of research fatigue and research resistance (Clark 2008).

Though Palestinian fieldworkers and academics upheld a strong political consciousness for social justice, their independence and values may be compromised by these demobilising factors to the extent that Arda and Banerjee recorded interviews where PNGO workers would admit to 'accepting irrelevant projects' in order to attract donors and market themselves, even when unqualified to do so (Arda and Banerjee 2019). This potential for project duplication and pointless research demonstrates, based on a profit–loss approach, an inherent aim to maximise profits without accounting to the ethics behind, going to the extent that the utility behind an NGOisation analysis is its critical stance on NGO accountability: 'who are the NGOs accountable for: those who it serves or those who provide its funding?' (Rodgers 2019, 79) which easily translates into a framework for the accountability of research, a crucial question in an over-research analysis. Over-research then occurs because PNGOs are located in close proximity and have a high acceptance ratio of projects.

To sum up, Islah Jad has argued that NGOisation encouraged a class-based movement, where NGOs became career catalysts for educated middle-class Palestinians (Hajjar 2001). In the context of Palestine, it is described as the projectisation of peace, where conflict resolution becomes 'a power base for the NGO elite to reach decision-making positions' (Jad 2007, 626) while rendering 'target groups' like the Bedouins and Palestinian scholars disenchanted with research.

Burn-out exhaustion among Palestinian scholars

Research in the West Bank is intrinsically nurtured by the politics surrounding the Palestinian–Israeli conflict. Any claim of over-research in the oPts is, therefore, context dependent. In a broader level, fieldwork in conflict zones, whether pro-Zionist, pro-Palestinian, or allegedly objective, inherently becomes entangled in a polarised field of biases. The '*researcher*', objectively a disruptive participant, probes and engages within these dialectics of conflict and 'truth discourses'. On one hand, research is associated with a core value of progress and development, and on the other, connected with interference, exploitation, and resistance. Therefore, institutionalised research has crucial effects on civil society as a knowledge producer, and critically, on the beneficiaries/recipients of research projects. Research, as an 'objective entity', has the potential to (re)produce biases, endorsing behaviours and beliefs.

The deciding factor that grounded my analysis on research ethics was when my Palestinian mentor suggested to '*make up my interviews*' (personal communication with Palestinian researcher, July 29, 2019). This advice was offered after I voiced my concerns over my fieldwork and my inability to gather interviews. At first, I assumed that my mentor was joking, however, to '*make up the interviews*', which reminded me my first day when he stated as a matter of fact that '*everything has already been written*' and that I should reword and recycle what has already been said about the Palestinian–Israeli conflict. He suggested: '*how about you pretend I am a Bedouin and you interview me? That will enrich your interview sample.*' When I declined, he sarcastically replied: '*you are too ethical*'.

'*You are too ethical*'. This problematic statement accompanied me throughout my fieldwork, leading me to believe that there was something inherently wrong with research. It made me wonder who else was making up interviews and gaining reputation from data fabrication. Dissatisfied with the research ethics of the non-profit, I reached a breaking point and became disenchanted with my contribution to the field.

It became apparent, after leaving the field, that the burden of conflict takes a toll on committed researchers. Just as I had reached a breaking point around my second month, my mentor had done so long ago. An accumulation of burdens: the first Intifada, the Oslo Accords, the second Intifada, and so on. Promising resolutions and hard-knock disappointments. At some point, through the professionalisation of the Palestinian national struggle and foreign encroachment, the disillusionment with research had co-opted any progress for resolution. The emotional exposure to the conflict's injustices had a burn-out effect that correlates closely to an exhaustion with research. The inevitability of losing a homeland leads to an emotional strain on Palestinian researchers. Insofar, that this has been described as a: 'practiced tension, a cautious anticipation of what would come next, a practical look to an already limited future from the viewpoint of a history of injustice, arbitrariness, and scapegoating' (Rothenberg 2016, 27).

Therefore, the futility of research is intertwined within these burn out attitudes amongst Palestinian scholars. As Arundhati Roy states, 'NGOs have funds that can employ local people who might otherwise be activists in resistance movements but now can feel they are doing some immediate, creative good (and earning a living while they're at it' (Roy, 2016, 335). Though a priori NGOs may be perceived as beneficial, they also serve as policing agents. Therefore, Palestinians who might engage in resisting the occupation become reliant to upkeep their income through donors. This has a demobilising effect on popular struggles. Adam Hanieh, referencing Franz Fanon, stated that settler colonialism is based on methods of domination and power exercised in most aspects of Palestinians' everyday life (Hanieh 2016, 38). Development's futility, Hanieh describes, is based on its illusory promotion of 'empowerment without power' (*ibid.*, p. 33). These power dynamics associate directly to Israeli colonial domination as individuals (gatekeepers) view themselves as trapped against the wall of Israeli colonisation cornered into seemingly limited alternatives, and therefore, are compelled to follow one option: compliancy with Israel's agenda. As such, they perceive their situation as hopeless. In particular, when tied to colonialisation as it takes part on the domination of the self, engaging in the dismantling of 'feelings of self-worth and self-esteem—and this, it is important to emphasize, does not refer solely to an individual sense but very much a collective spirit—it becomes difficult to believe in, conceive of, and struggle for different alternatives' (*ibid.*, p. 36).

As I interviewed Palestinian academics and potential gatekeepers, I encountered a folklorist who stated that: 'you know, I have many keys, when I do research, they give me the keys. These are their keys' (interview with Palestinian academic, August 12, 2019) showing around the room the old keys of Palestinian houses. This statement surprised me as the keys of return of Palestinians are proof of their right to return. However, it demonstrated the dynamics of research: the giving nature of the researched, their trust and personal stories, and the taking nature of research. I recognised how researchers sought out keys, the fetishism of truth illustrated in Westernman's lyrics, which leads to over-research fields as word to mouth spreads, attracting more researchers.

Accessibility and the recruitment of same participants

> Today, a local shopkeeper, drove me to a Bedouin community. Before we arrived, he started describing how he is used to researchers, the way we speak, the questions, and the answers. He said, 'I once brought another Finish woman here before. I promise you. I can guess your questions' which only added to my uneasiness. I know for a fact that the constant recruitment of the same participants disembogues in research fatigue.
>
> (Author's fieldwork diary, July 27, 2019)

This fieldwork diary entry in my first month expresses my constraints in the field. I will position the over-research issue to the easy accessibility to the West Bank, following Elisa Pasccuci's (2016) analysis where she linked high level of researchers to the humanitarian infrastructure's 'multi-layered network of outsourcing' (Pascucci 2016, 252). Koen et al. (2017) record Pascucci's correlation between convenience and accessibility to over-researched communities, when a researcher is quoted stating:

> Although not at the expense of justice, certain convenience factors, such as geographic accessibility, were seen as legitimate considerations in making selection decisions to ensure timely and cost-effective completion of research.
> (Koen et al. 2017, 5)

Therefore, the current easy accessibility to the West Bank plays a key factor in the saturation of researchers in the area. Though long-term onsite research in the West Bank is difficult to obtain due to Israel's visa denials, short-term research is easier as researchers can disguise themselves as tourists entering the West Bank. PNGOs are aware of this, and so, the research institute advised me to conceal the purpose of my visit as tourism. Israel allows tourist visitors a maximum of six months per year before raising concerns.

Most research institutes in the West Bank are located in humanitarian NGO hotspots in Area A which increases the reliance on these humanitarian gatekeepers to attract researchers. As Pascucci (2016) reports, what attracts researchers to a certain hotspot is security issues, a stable and accessible government, and popular issues within academia. The tragedy of the Palestinian–Israeli conflict, as mentioned earlier, has always been a popular subject in academia. Sukarieh and Tannock (2012), state that over-research goes side by side with under-research. They point out that shark-like academics 'follow the blood trail', linking research interests with macabre events and giving the example of the sudden interest of the under-researched Nahr El Bared refugee camp after its destruction in 2007 (Sukarieh and Tannock 2012). Mona Abaza (2011) would describe a similar trend of 'overnight Middle Eastern experts' who would invade Cairo searching *the authentic* during the Arab Spring. Edward Said described these academic works as disguised under 'the clothing of scholarship' (Said 1993, 314). This is corroborated in Abaza's grievances about 'Academic tourists sightseeing the Arab Spring' which illustrates the asymmetric relationship between 'North'/'South' academics, as western researchers 'typically make out of no more than a week's stay in Cairo . . . to tag themselves with the legitimacy and expertise of first-hand knowledge' (Abaza 2011). While I was at the institute, I experienced first-hand foreign visitors, most of them young students, like me, eager to go into the field and contribute to the conflict's resolution. A Palestinian colleague told me twice that my stay of three months 'was not enough. Minimum should be a six-month stay' (personal communication with a Palestinian researcher, July 21, 2019) which I could not comply with, as it conflicted with my last two semesters at the university. Though the research institute limited research volunteering to a minimum of three months, this policy was not followed strictly. The director accepted visiting foreign students staying as little as one week. Despite my short stay of three months, objectively, I was the longest visiting student researcher. Therefore, though policies may be set in place to limit the number of visiting scholars and protect Palestinian communities, these are not followed strictly, allowing a profiting system of gatekeeper tour guides.

The gatekeeper as a tour guide metaphor is useful as it illustrates the controversial notions of exploitation versus local investment. It is also eerily similar to M.G. Khan's (2015) analysis of gatekeepers 'pimping' to researchers, in consideration to his experience as a youth worker. As Pascucci (2016) notes, gatekeepers have the capacity to ward off communities from participating in projects, explaining how once the Jordanian humanitarian organisations decided to oversee and filter beneficial projects from counterproductive ones, the researchers and organisations would move to regional countries with less-restrictive protocols, such as the West Bank. This entails that researchers are aware of the factors fuelling the over-research issue. Yet, the issue is ignored as a matter of practical and profitable research for the researcher's benefits.

Therefore, Khan stresses that gatekeepers must uphold the important responsibility as the middlemen between researchers and their communities to protect them from harm by screening 'friendly' researchers seeking to organize visits, focus groups, surveys, interviews, etc. (Khan 2015). I underwent several of these credentialisation screenings during my fieldwork, where I would meet potential gatekeepers working with Bedouins and undergo an interview where I would try to convince them of my well-founded intentions of my research. Gatekeepers, however, as Khan elucidates, are used to these sympathetic researchers who declare an open-mind and 'good faith' intentions. This screening becomes critical for gatekeepers to ascertain newcomer's objectives and affiliations in the conflict, as they could be surveillance agents, potentially putting the Bedouin communities in harm's way (informal interview with Bedouin activist, August 17, 2019). Therefore, gatekeepers have to be good readers of character when introducing researchers to their communities (*ibid.*). I was advised by a Bedouin activist that releasing sensitive data could expose the Bedouins to be considered as Israeli collaborators. Therefore, a well-intended research project can give the means and legitimisation for Israeli agencies to facilitate their displacement.

Futility of research: the issue of reciprocity and forms of practised speech

Conceptually speaking, over-research works as a form of stagnant circulation of knowledge as the project's beneficiaries and the research's audience are disassociated from each other. Where contributions circulate in inaccessible 'elite' circles, shrouded in highly technical and academic jargons. Therefore, common frustrations relate to issues of reciprocity as researchers tend to stay in their ivory towers (Browne and Moffett 2014). This encourages research resistance as the given trust of the participant has been broken, propelling the expectation in researched communities that the research outcome will not be distributed among their village.

While conducting research, not once was I asked when and where they would be able to read the outcome of my research. Perhaps, due to my lack of credibility and status but more likely because few researchers come back to disseminate their findings (Khan 2015).

I will now argue that the lack of knowledge exchange in an over-research field foments forms of *practised speech*. A practised speech refers to informants answering interview questions as if reciting from a memorised text. It implies an over-interviewed data sample, displaying signs of research fatigue. This is characteristic of conflict research as individuals seeking aid must recite their plight in order to receive assistance. It entails a practised narrative of conflict. The implication is that:

> research fatigue could potentially create distorted, invalid results because either the participants respond in a learned way, or in ways that misrepresent reality because they no longer take the study seriously.
>
> (Koen et al. 2017, 4)

The practised speech phenomenon was noticeable when deviating a question from the standardised surveys most Palestinians are used to. Common interview questions include: *How does the occupation affect you? How do you feel about it? Do you have hopes for a two-state solution?* These questions are continuously asked due to relentless pilot studies, field 'newbies', and tourists. This was visible with older generations refusing to participate in research while younger Palestinians were seemingly more open to its contribution. During an interview with a young, English-speaking, Bedouin man, his mother was present throughout the whole process, silent except while offering us tea and coffee. As the son explained on his mother's non-participation: 'if I will translate it to her and ask her what's peace, what's justice, she will kick you and me out' (interview with young Bedouin man, July 27, 2019). The explanation implied that research fatigue correlates to constant questioning on the emotions of peace-resolution, causing fatigued narratives of conflict. The Bedouin man would bid farewell, stating that to speak of justice, when nothing is done for justice, is to speak in barren words (informal interview with young Bedouin man, July 27, 2019).

In an email exchange after I thanked him for the interview, he would politely let me know that 'it was a good interview, but I'm looking behind the interviews, we want to get more people here and let them know the situation. You are welcome to come visit us again' (Email exchange with Bedouin man, July 31, 2019). Interviews, therefore, do not provide concrete solutions, and instead, the village representatives seek international advocacy with visitors helping and engaging in their community development projects.

Burden of conflict: the issue of research *savviness* and research scams

A continually researched community eventually develops a nurtured familiarity with research. Pascucci has defined it as 'research savviness' (Pascucci 2016, 251) which entails that researched communities have solid knowledge of the ethical guidelines on how research has to be conducted, hopefully regulating researchers' unethical practices and attitudes. However, the risk of an overly 'research savviness' community in a neoliberalist environment, I will argue, often accompanies researchers being taken advantage of.

A prominent by-product of NGOisation and the over-research issue is the emergence of research scams in the West Bank. These are 'NGOs' who benefit from the surplus of researchers and demand for participants, preying on researchers unfamiliar with the field and desperate for interviews. I encountered this situation while chatting with an activist about possible contacts that I could be referred to. During the conversation, I mentioned the name of the scam NGO as one of the few NGOs that replied to my email. This specific research NGO offered free accommodation in the Jordan Valley. However, as the activist would warn me, once the researcher arrived, the total price of the food and transportation fee would increase each day, and the researcher would be compelled to pay due to the remoteness of the location (informal interview with Bedouin activist, August 17, 2019). Therefore, my advice in an over-researched place is to be cautious of willing gatekeepers as they are well-aware of the high demand for their services and the alluring funding awarded to western researchers. In these competitive settings, it is important to find trustworthy gatekeepers who can direct you in the right direction.

Thus, research savviness is a concept that can be extended to researchers who must first familiarize themselves with their surroundings and develop awareness skills to navigate the field. Whether it is avoiding research scams and/or dangerous encounters, the illustrative was when a trusted gatekeeper admits that their NGO would not dare to venture into every Bedouin village because of certain individuals with a strong, even aggressive, resistance to researchers (informal interview with Bedouin activist, August 17, 2019).

Conclusion

In evaluating the extent to which NGOisation affects qualitative research in the West Bank, this chapter analyses the factors and root causes of how the Israeli–Palestinian conflict became an over-researched place. Its implications will affect the researcher's accessibility to the field, as one must be skilful in navigating the politics of research and the increasing competition between different NGOs; now acting as gatekeepers. However, at its core, it becomes a question of gaining trust from over-researched communities and over-zealous gatekeepers. It is key to resolve the tensions associated with over-research's exploitative nature. The applicability of this critical analysis is to maintain a moral stance in order to 'keep fieldwork dialogically alive' (Conquergood 1985, 10), which entails opening dialogue to criticism and amendment.

Browne and Moffett (2014) make an important point that most researchers struggle in the field because fieldwork is a 'learn as you do', which is something early-career researchers are pushed towards without strong training at the university level. This system of learning by doing is, most importantly, negative for vulnerable communities who have become training material; part of the researcher's learning curb of trial and error.

As such, Browne and Moffett (2014) describe a 'culture of silence' regarding the hardship of fieldwork in these closed-context research settings. The culture of silence in the West Bank surrounds the NGO business model, and consequently, shrouds the over-research phenomenon. Over-research occurs due to this enabling system as the researchers are ill-prepared to face the difficulties of research: the emotional toll of conflict, confidentiality dilemmas, forcing research on over-researched participants, and preserving our core ethical values on the face of a research field that rejects you. Over-research in the West Bank becomes a contradiction between pushing researchers to become armchair researchers while encouraging their contribution to over-researched issues. My point here is to stress the importance of breaking the silence of the root causes of over-research in the West Bank. As for today, the implications of doing research in the West Bank are hard on the researcher and researched alike. It is my belief that research has reached a standstill, which foments perceptions of over-research. Perhaps it becomes obvious that my Palestinian mentor considered fieldwork, with its long interviews and repetitive questions, as everyday life intrusions. Nevertheless, ethical protocols, such as First Nation's OCAP (Ownership, Control, Access, and Possession) (Schnarch 2004), are being used by some to address research's structural issues. Until structural reforms are implemented to improve NGOs and researchers' attitudes towards the welfare of their participants, perhaps the most ethical step would be to halt further research.

References

Abaza, Mona (2011) "Academic Tourists Sight-Seeing the Arab Spring." *Jadaliyya* (blog). Accessible from www.jadaliyya.com/Details/24454

Adams, Tony E., Carolyn Ellis, and Stacy Holman Jones (2015) *Autoethnography*. Oxford: Oxford University Press.

Amara, A., and M. Nasasra (2015) "Bedouin Rights under Occupation: International Humanitarian Law and Indigenous Rights for Palestinian Bedouin in the West Bank." *Norwegian Refugee Council (NRC)*. Accessible from www.nrc.no/globalassets/pdf/reports/bedouin- rights-under-occupation.pdf

Arda, Lama, and Subhabrata Bobby Banerjee (2019) "Governance in Areas of Limited Statehood: The NGOization of Palestine." *Business & Society*, 000765031987082. https://doi.org/10.1177/0007650319870825.

Browne, Brendan C. (2018) "Writing the Wrongs: Keeping Diaries and Reflective Practice." In *Experiences in Researching Conflict and Violence*, edited by Althea-Maria Rivas and Brendan Ciarán Browne, 187–203. Bristol: Bristol University Press, Policy Press.

Browne, Brendan C., and Luke Moffett (2014) "Finding Your Feet in the Field: Critical Reflections of Early Career Research on Field Research in Transitional Societies." *Journal of Human Rights Practice* 6 (2): 223–237. https://doi.org/10.1093/jhuman/huu010

Choudry, A. A., and Dip Kapoor (2013) *NGOization: Complicity, Contradictions and Prospects*. London: Zed Books.

Clark, Tom (2008) "'We're Over-Researched Here!': Exploring Accounts of Research Fatigue within Qualitative Research Engagements." *Sociology* 42 (5): 953–970. https://doi.org/10.1177/0038038508094573.

Conquergood, Dwight (1985) "Performing as a Moral Act: Ethical Dimensions of the Ethnography of Performance." *Literature in Performance* 5 (2): 1–13. https://doi.org/10.1080/10462938509391578.

Cooley, Alexander, and James Ron (2002) "The NGO Scramble: Organizational Insecurity and the Political Economy of Transnational Action." *International Security* 27 (1): 5–39. https://doi.org/10.1162/016228802320231217.

Deloria, Vine (1969) *Custer Died for Your Sins: An Indian Manifesto*. Norman: University Of Oklahoma Press.

De Voir, Joseph, and Alaa Tartir (2009) "Tracking External Donor Funding to Palestinian Non-governmental Organizations in the West Bank and Gaza Strip 1999–2008." *Palestine Economic Policy Research Institute (MAS)*. Accessible from http://eprints.lse.ac.uk/50311/

Ferguson, James (2014) *The Anti-Politics Machine "Development", Depoliticization, and Bureaucratic Power in Lesotho*. Minneapolis: University of Minnesota Press, 1990.

Goffman, Erving (1956) *The Presentation of Self in Everyday Life*. Edinburgh: University of Edinburgh Social Sciences Research Center.

Hajjar, Lisa (2001) "Human Rights in Israel/Palestine: The History and Politics of a Movement." *Journal of Palestine Studies* 30 (4): 21–38. https://doi.org/10.1525/jps.2001.30.4.21.

Hanieh, Adam (2016) "Development as Struggle: Confronting the Reality of Power in Palestine." *Journal of Palestine Studies* 45 (4): 32–47. https://doi.org/10.1525/jps.2016.45.4.32.

Jad, Islah (2007) "NGOs: Between Buzzwords and Social Movements." *Development in Practice* 17 (4–5): 622–629. https://doi.org/10.1080/09614520701469781.

Jellinek, Lea (2003) "Collapsing under the Weight of Success: An NGO in Jakarta." *Environment and Urbanization* 15 (1): 171–179. https://doi.org/10.1630/095624703101286439.

Khalidi, Rashid (1997) *Palestinian Identity: The Construction of Modern National Consciousness*. New York: Columbia University Press.

Khan, M.G. (2015) "Anthros and Pimps Doing the God Trick: Researching Muslim Young People." In *Research and Policy in Ethics Relations: Compromised Dynamics in a Neoliberal Era*, edited by Charles Husband, 105–121. Bristol: Bristol University Press, Policy Press.

Khun, Thomas (1970) "The Structure of Scientific Revolutions." *Poscript* 69.

Koch, Natalie (2013) "Introduction—Field Methods in 'Closed Contexts': Undertaking Research in Authoritarian States and Places." *Area* 45 (4): 390–395. https://doi.org/10.1111/area.12044.

Koen, Jennifer, Douglas Wassenaar, and Nicole Mamotte (2017) "The 'Over-Researched Community': An Ethics Analysis of Stakeholder Views at Two South African HIV Prevention Research Sites." *Social Science & Medicine* 194: 1–9. https://doi.org/10.1016/j.socscimed.2017.10.005.

Kusenbach, Margarethe (2003) "Street Phenomenology." *Ethnography* 4 (3): 455–485. https://doi.org/10.1177/146613810343007.

MAS Palestine Economic Policy Research Institute research team. Rep. (2007) *Mapping Palestinian Non-Governamental Organizations in the West Bank and the Gaza Strip*. Jerusalem and Ramallah: MAS Palestine Economic Policy Research Institute.

McDowell, Sara, and Máire Braniff (2014) "An Intractable Conflict and an Irreconcilable Past: Contesting the 'Other' through Commemoration in Israel/Palestine." In *Commemoration as Conflict. Rethinking Peace and Conflict Studies*. London: Palgrave Macmillan. https://doi.org/10.1057/9781137314857_7

Neal, Sarah, Giles Mohan, Allan Cochrane, and Katy Bennett (2016) "'You Can't Move in Hackney without Bumping into an Anthropologist': Why Certain Places Attract Research Attention." *Qualitative Research* 16 (5): 491–507. https://doi.org/10.1177/1468794115596217

Pappé, Ilan (2006) *The Ethnic Cleansing of Palestine*. Oxford: Oneworld Oxford.

Pascucci, Elisa (2016) "The Humanitarian Infrastructure and the Question of over-Research: Reflections on Fieldwork in the Refugee Crises in the Middle East and North Africa." Royal Geographical Society (with IBG). John Wiley & Sons, Ltd. Accessible from https://rgs-ibg.onlinelibrary.wiley.com/doi/full/10.1111/area.12312.

Rodgers, Carlyn (2019) "The NGOization of Pro-Black Mobilization in Rio de Janeiro: An Investigation of Agency between Pro-Black Activists and Anistia International Brasil." *Fletcher Forum of World Affairs* 43 (2): 73–94.

Rothenberg, Celia E. (2016) *On Doing Fieldwork in Palestine: Advice, Fieldnotes, and Other Thoughts*. London: Palgrave Macmillan.

Roy, Arundhati (2016) *The End of Imagination*. Chicago: Haymarket Books.

Said, Edward W. (1993) *Culture and Imperialism*. London: Vintage.

Schnarch, Brian (2004) "Ownership, Control, Access, and Possession (OCAP) or Self Determination Applied to Research: A Critical Analysis of Contemporary First Nations Research and Some Options for First Nations Communities." *Journal of Aboriginal Health*, 80–95.

Sukarieh, Mayssoun, and Stuart Tannock (2012) "On the Problem of Over-Researched Communities: The Case of the Shatila Palestinian Refugee Camp in Lebanon." *Sociology* 47 (3): 494–508. https://doi.org/10.1177/0038038512448567.

Verma, Gita Dewan (2002) *Slumming India: A Chronicle of Slums and Their Saviours*. New Delhi: Penguin Books.

Whitaker, Urban (1963) "The Dangers of Over-Research." *Background* 6 (4): 65. https://doi.org/10.2307/3013632.

7 Overlooked cities and under-researched Bharatpur, Nepal

Hanna A. Ruszczyk

Introduction

While this book is primarily concerned with over-researched places and the implications that research-density has on the people and places researched, on the researchers, on knowledge produced, and on the theories that are developed, the purpose of this chapter is different. This chapter is introducing the concept of academically overlooked cities and what is lost by ignoring them. The organisation of this chapter is as follows: a theoretical framing of overlooked cities is presented, followed by a description of the *Overlooked Cities: Power, Politics, and Knowledge Beyond the Urban South* Book (2021), and then an overview of an overlooked urbanising country (Nepal) and an under-researched city (Bharatpur) is presented.

Theoretically overlooked cities

Some cities and urban processes always seem to be kept out of view and remain removed from familiar ways of seeing, questioning, and engaging. And yet, these cities are still subjected to the various assumptions made and the categories and labels devised by urban and human geography scholars and also by international and national policymakers. Critical urbanists have long emphasised the partiality of urban theory, demonstrating how patterns of urban knowledge production reflect *particular* historical, institutional, political, economic, and cultural formations. Urban knowledge travels between cities and across continents and, in the process, it may be argued, becomes 'richly populated in place, region, networks, and in conversation' (Oldfield 2014, p. 7). Contemporary urban knowledge is based on specific spaces while ignoring most cities throughout the world.

Through the use of the concept of overlooking, it is possible to explore the dynamics of 'overlooked-ness' and the consequences of overlooking as they are manifested in particular cities and in specific contexts. It is important neither to continue the academic confinement of these cities to the margins nor to only mainstream these until-now marginalised cities into existing urban theory. These cities are different from the capital, mega, or global large cities. The relationships between residents and local authorities, the fluidity between rural and urban

DOI: 10.4324/9781003099291-8

spaces and also the knowledge that travels from international to local spaces due to migration are different in these cities. Insights from over-researched cities do not suit these cities and do not sit well.

While overlooking a city may marginalise it from important international and national policy discussions and resource allocations, it may simultaneously provide local actors with the space to experiment, create, and innovate without worrying what others will think and how they will react. In addition, it is important not to fetishise or romanticise the ordinary nor to be content with 'ordinariness' as a satisfactory analytical category. Rather it is important to challenge how the accumulation of urban knowledge has meant that certain cities—particularly in the global North but also the megacities and global cities in the South—have received special attention in academic literature and policymaking.

Within critical urban theory, there remains an overwhelming focus on the world's largest cities. This is not merely a case of under-representation. Paterson et al. (2017, p. 109) explain that smaller cities face disproportionate risks not only due to the concentration of 'most of the world's vulnerable urban populations' in these cities but also due to the 'limited data, political power, personnel, and resources'. Jorge Enrique Hardoy and David Satterthwaite have dedicated much of their academic careers to these kinds of cities, and their work has provided the foundational thinking behind the orientation of research on overlooked cities. It is hard to believe that it was 35 years ago, in *Small and Intermediate Urban Centres*, Hardoy and Satterthwaite (1986) first argued that urban studies lacked a detailed and nuanced account of the diversity of urban centres and how smaller cities interacted with their surrounding rural areas.

Subsequently, in their 1989 book, *Squatter Citizen*, because Hardoy and Satterthwaite recognised that many of the largest Southern cities were developed by colonial powers and thus subjected to their peculiarities, Hardoy and Satterthwaite focused their attention on the specificities of life in smaller cities. To do so, Hardoy and Satterthwaite argued, it requires an understanding of their 'own unique mix of resources, development potential, skills, constrains and links with the surrounds and the wider regional and national economy' (1989, p. 299). They were arguing for a careful, nuanced understanding of smaller cities. They were trying to push back against the glorification of large, mega over-researched cities.

Reflecting on the book three decades later, Satterthwaite reiterated the urgency for this line of enquiry, explaining how 'the rights [of residents] to water, education and healthcare are often denied' in these overlooked cities (2019). Hardoy and Satterthwaite's commitment to the urban populations living in these cities warrants further attention and care in order to create 'a better body of knowledge [to] widen the understanding of small and intermediate urban centres' (1986, p. xix). More scholarship is required to serve these cities with justice, compassion, and insight. This chapter is a small endeavour in this direction.

Following the publication of these two books in the 1980s, postcolonial urbanism views the constitution of the urban as 'always variable, polymorphic and historically determinate' (Robinson 2014, p. 62) and thus seeks to disrupt the universalising tendencies of urban theory (and the over-research of some cities who shall not be named) through conceptual innovation and the examination of alternative urban geographies. This is not merely an intriguing intellectual exercise but has a political and moral imperative, given its potential to rewrite urban knowledge to account for a much larger proportion of the world's urban population.

Southern and postcolonial urbanisms have been enormously influential in refocusing attention towards cities throughout the Global South, and yet many cities continue to be systematically overlooked. While megacities and capital cities function as 'city states in a networked global economy, increasingly independent of regional and national mediation', other cities are left to 'seek new ways of claiming space and voice' (Appadurai 2001, p. 25). The lack of attention given to smaller cities (Ali & Rieker 2008, p. 2) is a self-imposed limitation on our understanding of the urban cities and it implies that these cities are less worthy of critical analysis or that they experience the same urban development issues but on a different scale (Sheppard et al. 2013, p. 894). Methodological, theoretical, and conceptual frameworks have yet to position smaller and/or more regional cities in the front and centre.

Overlooked cities: power, politics, and knowledge beyond the urban South

This section provides an overview of a recently published book that presents insights about theory and empirical work arising from overlooked cities. The *Overlooked Cities: Power, Politics and Knowledge Beyond the Urban South* book (Ruszczyk et al. 2021a) was published by Routledge in 2021. To overlook is not merely to ignore. It may involve a conscious choice to look elsewhere or it may constitute an act of simultaneously knowing but not caring. Processes of overlooking strike at the very heart of what we understand to be 'the urban' throughout many parts of the world.

The book *Overlooked Cities* reflects on and impacts the changing landscape of urban studies and geography from the perspective of smaller and more regional cities in and beyond the conventional understandings of the urban South. The *Overlooked Cities*' collection of essays focuses on 13 cities in nine countries and across three continents: Luzhou, China; Bharatpur, Nepal; Bloemfontein/Mangaung and Pretoria/Tshwane, South Africa; Zarqa, Jordan; Santa Fe, Argentina; Manizales, Colombia; Arequipa and Trujillo, Peru; Dili, Timor-Leste; Bandar Lampung, Semarang, and Bontang, Indonesia. The book was written by early-career researchers, each of whom focused on regional or smaller cities overlooked by urbanists. All the researchers struggled to find urban theory that spoke to the empirical settings.

The *Overlooked Cities*' attentiveness towards the dynamics and processes of overlooking allows us to critically examine the ways in which cities are uniquely positioned within different urban and knowledge hierarchies. In many respects, these asymmetries define the various aspects of urban condition in these cities and are reflected in matters of governance, urban planning, disaster risk reduction, climate change adaptation, international development assistance, and the activities of non-governmental organisations.

The book *Overlooked Cities* aims to (1) unpack the dynamics of 'overlookedness' in these cities, (2) identify emerging trends and processes that characterise such cities, and (3) provide alternative sites for comparative urban theory. It is organised into two themes: firstly, politics and power, and secondly, production and negotiation of knowledge. The authors share a commitment to challenging the unevenness of urban knowledge production by approaching these cities on their own terms. Only then can we as scholars harness the insights emanating from these overlooked cities and contribute to a deeper and richer understanding of the urban itself. These cities are special due to their ordinariness (Robinson 2006). The sheer number of such cities in the world matters to inform different practices of ordinariness in a heterogeneity of place and context. Their ordinariness and commonness make them simultaneously situated in an ambiguous situation of importance and unimportance. There is something different about them but also something conventional.

The *Overlooked Cities* is a *political* and *critical* project. It is vital in this century where most people do live in non-megacities, that we research smaller, regional, secondary, tertiary, and intermediate cities and understand what is taking place. Why should some sites be over-researched and where new knowledge is incremental at best when there are countless cities that have not been researched? There is a political and critical imperative at stake. The book *Overlooked Cities* actively challenges and questions power structure and relations not only in the production of cities but also in the production and negotiation of urban theory. It is a complex and contested urban space. As a political project, it is double-handed in its approach. There is something significant in the silence and omissions where we find these cities. As an intellectual and critical project, we consider overlookedness as an opportunity to critically engage with the existing urban theory and to find new names and terms that can form the foundation of a new agenda for theoretical engagements with cities and the urban throughout the world.

An overlooked urbanising country and city (Nepal and Bharatpur)

In this section, an example of an overlooked and under-researched urbanising country and a city within that country are described. Nepal, sandwiched between the political and economic behemoths of India and China, is a country of almost 30 million residents. In the recent past, Nepal has survived a 10-year internal conflict (1996–2006) that had killed 13,000 people, an earthquake in 2015 that killed almost 9000 people, and experiences an everyday economic and social landscape where increasing international remittances from young men are the backbone of the economy (Ruszczyk 2017).

To give a bit of historical context, this formerly secluded Hindu kingdom opened its boundaries to the outside world in 1951 (Toffin 2013). In the 1990s, there was multiparty democracy, and in 1992, a newly elected government introduced legislation that provided for powers and responsibilities of local governments to be expanded (Acharya 2018). In the midst of the Maoist Insurgency

(1996–2006) (also known as the Maoist People's War), in 2002, the government led by the king replaced locally elected officials with central government employees because the Maoists were gaining control of the local governments. Since 2002, this lack of elected representation at the municipal and ward levels created a situation where the ability of residents to influence the local authority has been curtailed. In 2006, a comprehensive peace accord was signed and a republic was established in 2008. From 2008 to 2015, a new constitution was debated without any progress. To date, the government of Nepal struggles to provide basic services for its citizens (such as electricity and solid waste management) but is simultaneously able to govern, often through informal mechanisms (Nightingale et al. 2018; Ruszczyk 2017).

Newly urban Nepal

In 2001, there were only 58 municipalities (with a population of 10,000 or more) including the metropolitan city of Kathmandu and four sub-metropolitan cities (Tanaka 2009). The country was overwhelmingly rural with 86% of the population living in rural areas (IFAD 2014) and over 40% of the population needed to walk two hours to access a paved road. The country had a centralised form of government and minimal autonomy for centrally appointed government officials working in municipalities.

In 2014, due to power struggles between the Ministry of Federal Affairs and Local Development (MoFALD) and the Ministry of Urban Planning regarding who would manage the rapidly urbanising cities of Nepal, the central government administratively transformed Nepal. The central government incentivised rural government units called Village Development Councils to amalgamate into larger entities or to be absorbed by adjacent municipalities. MoFALD won the power struggle. In late 2014, there was an administrative leap to 191 municipalities.

By the autumn of 2015, the constitution was promulgated and a federal system was created with three levels of government: national, provincial, and local. There are now seven provinces and 753 local governments that comprised six metropolitan cities, 11 sub-metropolitan cities, 276 urban municipalities (*Nagarpalika* in Nepalese language), and 460 rural municipalities (*Guanpalika* in Nepalese language). By 2017, 60% of the population lived in urban municipalities (ADRA Nepal 2018). In these new urban spaces, municipalities and the urban overlap in ways that obfuscate what is happening in relation to infrastructure as well as who and what is being governed in Nepal (Ruszczyk 2020).

In 2017, the national government introduced the Local Government Operation Act 2074 with the goal of strengthening management and operational capacity (Acharya 2018). Local governments have now received a wide range of powers from the Local Government Operation Act of 2017, including local planning, economic development, social development, environment development, and community infrastructure. Understandably, the local governments are struggling to cope with the new and increased responsibilities (Asia Foundation 2018) and a lack of knowledge as to how to be urban (Ruszczyk 2020).

Bharatpur

Bharatpur, a metropolitan and regional city, is located on the fertile plains of Nepal near its border with India. It has developed as the main economic and social hub of the central region of Nepal. Bharatpur is the fourth most populated city of Nepal (280,000 residents) and second largest city in geographic area. Until November 2014, Bharatpur Municipality had 14 wards and a population of 144,000. By December 2014, it became a sub-metropolitan city (SMC) with 29 wards and its physical area increased by 50% and its population increased by over 50% (to 210,000) due to the amalgamation of five villages in the southeast and southwest of the city. These villages brought their rural poverty and specific hazards (river flooding and wild animal attacks from the jungle) to the newly created SMC. Five months later, an earthquake struck and the local government struggled to inspect the damaged 8% of 40,000 buildings.

In 2017, Bharatpur became a metropolitan city and its geographic area increased again and its population increased to 280,000 residents due to the addition of adjacent villages. These changes have made Bharatpur very rural in terms of demographics, geographic coverage, and physical attributes such as quality of roads, access to electricity and piped water, and access to sanitation. At the same time, municipal elections had taken place in the Spring of 2017, under the new federal government system. Since the King replaced the elected officials in 2002, these were the first elections for local officials. The political landscape is changing in Nepal and this directly changes how the urban areas and cities are not only conceptualised but governed.

Bharatpur is on the Terai, the plains of Nepal, which have been heavily researched by anthropologists and cultural geographers who are interested in identity and political struggles taking place during the past decades (see Lawoti and Guneratne 2013; Chaudhary 2011; Guneratne 2002). This academic writing is also focused on the rights, representation, and belonging of the Indigenous population of Nepal. While half of Nepal's population resides in the Terai, there has been minimal urban research into the cities located there. This led to my research on Bharatpur, one of the five largest yet overlooked regional cities of Nepal (see Figure 7.1).

The radical and rapidly imposed administrative changes are straining the capacities of the newly elected local authorities to cope with the new and increased responsibilities (Asia Foundation 2018). For example, the newly elected mayor and municipal executive committee of one of the most important cities of Nepal, the Metropolitan City of Bharatpur, now leads a local government that has a significantly broadened range of responsibilities including education and hospitals. The economy of Bharatpur was traditionally dependent on agriculture but due to increased migration and economic activity, there are emerging changes—fertile agricultural land is being utilised for housing and industrial development.

A key function of the local authority is to consider long-term development of the city. Urban planning is a necessity. How it is being considered in a regional city may not be conventional. Urban planning is being considered in a careful, incremental manner radiating from the historically geographic centre, the urban

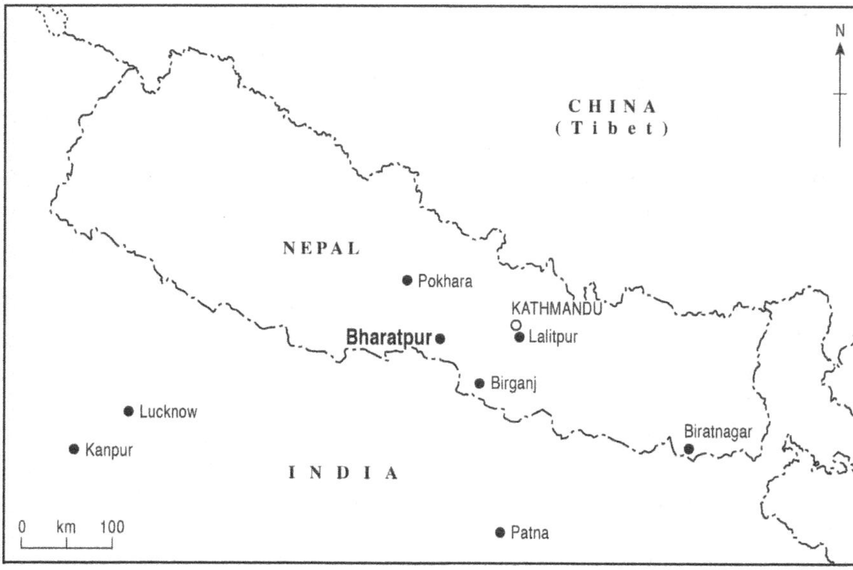

Figure 7.1 Nepal's largest cities (2019).

Source: Cartographic Unit, Department of Geography, Durham University (2019)

core, to the city's periphery, the rural areas (Ruszczyk 2020). The logic of urban planning is incremental because this is all the local government is capable of implementing at the present time, according to interviews with government officials of different levels and with residents. For the newly elected government officials, there is some attempt to utilise incremental urbanism in the hope to formalise, codify, and institutionalise elements of urban planning (Datta et al. 2019). The local authorities are incrementally learning how to implement the law and they understand the need to tread carefully and slowly in implementing urban planning regimes in a staggered manner to not upset the residents.

After the boundaries of Bharatpur were enlarged to amalgamate villages began in 2015, relationships between many residents and the local authority have not had the opportunity to be developed. It is in these geographic locations where the Metropolitan City of Bharatpur is incrementally deciding not to intervene, *yet* (Ruszczyk 2020). They are scared of upsetting the rural residents who pointedly ask the Metropolitan City of Bharatpur why they should follow rules they are not aware of and do not understand. They question why they should pay the local government a fee to get approval to build a house on their own land. The staff assigned to the ward offices (lowest level of local government) are insufficient in number and lack knowledge, social networks, and capacity to implement the law (Datta et al. 2019).

Often, Chitwan, Narayangarg, Narayangadh, and Bharatpur are used interchangeably to signify the same location. To highlight how problematic Bharatpur is to research, please consider this example of names and maps. Bharatpur is missing from maps of Nepal even though it is one of the four largest cities. It should be on maps but it is not. A quick google search of Bharatpur leads to a Wikipedia page (Wikipedia 2020) and on the map shown, the name of the city is not Bharatpur but Narayangadh. Narayangadh is the historical city centre which has existed for decades. Bharatpur was created as a municipality in 1979. Today, Bharatpur incorporates Narayanghadh. Bharatpur was the first municipality of Chitwan District. On other maps, the region which is called Chitwan is utilised instead of Bharatpur. When Nepali government officials and disaster experts utilised the name Chitwan in our meetings during 2013–2017, they were referring to a city they viewed as a political extension of Kathmandu's power in the Terai. To research a city, whose contemporary name is even called into question, becomes problematic. To name is to signify a point of view. Naming is power.

Historically, cities such as Bharatpur, Nepal have not been interesting to research. They are harder to reach (both physically and mentally) and they do not have the high voltage power of the capital. Ordinary, boring cities that have no particular claim to fame do not have researchers clamouring to research them. These cities are overlooked. Unlike megacities that have a certain prestige attached to them, medium-sized, regional cities of the urban South do not attract scholars.

Bharatpur reflects many characteristics associated with urbanising Nepal. Therefore, it should not only be more widely known but it should be researched more. It is a dynamic and heterogeneous city with a long-established population at its core, new affluent migrants building houses, migrants who fled conflict in their villages and towns in the nearby hill districts, new residents from nearby villages that are being amalgamated into the municipality, and economic migrants from Bihar State, India who explained that they earn more money in Bharatpur than at home in India. All of these residents have different connections with each other, with the government, and with the urban environment (physical, economic, social, and political). The newcomers to the city are heterogeneous; some assimilate rapidly due to their caste and extended family networks and finances. Others, who flee social tension and are forced to settle in Bharatpur without the extended family networks and finances must rent accommodations and find support systems more difficult to create.

I chose Bharatpur, Nepal as the empirical site for my 2013–2017 urban-based PhD research project (Ruszczyk 2017) after extensive consultation with national stakeholders and investigation regarding emerging urban issues in Nepal. In 2013, the urban signified Kathmandu Valley and Pokhara. Most other cities did not register as very important for knowledge creation and policy work on topics related to the urban such as urban planning, governance, or understanding urban risk and resilience. They were not on the radar of government officials nor academics. I found there was very little academic research on cities located outside of the Kathmandu Valley and the second largest city of Pokhara. National government officials, the donor community, international non-governmental organisations, and a Nepalese research organisation were very supportive of my decision to conduct research into risk perceptions and urban resilience in Bharatpur. They were keen to learn more about what I found in my research.

It never appealed to me to study a city or location that other foreign scholars researching the urban and/or cities had already written about. I did not want to debate with other academics in hotel dining rooms or in conference settings who had the more senior meeting during fieldwork or who had an 'adventure' in a particular informal settlement. I did not want to be influenced by what other foreign scholars thought about the city I was exploring. During my PhD research, I was on a quest to create academic 'new knowledge' on a city and contribute knowledge that would be of use to policy and practice. Of course, this is a naïve statement and a bit embarrassing to admit now, years later. What constitutes new knowledge? Does new knowledge only exist when it is written by a foreign scholar? Is it only new when it is published in a peer-reviewed academic journal behind a paywall that most people cannot access? (Clearly, the answer is no.)

Bharatpur offers opportunities to learn from its inhabitants: how people live and what aspirations they have for the future (Ruszczyk and Price 2019). This Nepali context is more similar to the context of urbanisation in the twenty-first century than the megacities and global cities. Navaro-Yashin (2012, xii) argues, 'That only certain spaces and themes make themselves available and accessible for study by certain people'. Navaro-Yashin's words rang true during my fieldwork and gave me hope when I faced doubts, struggles, and uncertainties. Analysing data and emerging findings arising from such a location when there are no common references or comparisons is difficult at times but, simultaneously, much insight can be *lost* when comparing to more visible and academically researched cities. I had to trust myself and what I heard from an overlooked, under-researched city and bring the views from Bharatpur's residents, government officials, and other stakeholders to the foreground of my academic and policy-facing writing.

Conclusion

There is an accumulation of urban knowledge based on certain over-researched cities—particularly in the global North but also the megacities and global cities in the South. This chapter has introduced the idea of academically overlooked cities and what is lost by ignoring them rather than addressing over-researched places. To overlook is not only to ignore but also has elements of not caring.

While this chapter's emphasis on researching the overlooked could be considered as only an intriguing intellectual exercise, researching overlooked cities is in reality a *political* and *moral* necessity. This type of research can contribute knowledge to account for places where the majority of the world's urban population resides rather than attempting to utilise theory and knowledge from one particular type of over-researched city and transporting it to the rest of the urban world.

The example of Bharatpur, Nepal reflects many characteristics associated with urbanisation and developing cities in the past decade. Cities such as Bharatpur should not only be more widely known but also researched more. Join the growing collective of early-career researchers from the book *Overlooked Cities* who are striving to make a space in urban theory for the spectacularly ordinary, boring cities where most of us live. For additional insights into overlooked cities, please look at the *Theorising From the Overlooked City* Magazine (Neves Alves and Ruszczyk 2021) and the *Overlooked Cities* website (Overlooked Cities 2021). Help us to make overlooked cities less overlooked by urban scholars.

Acknowledgements

Parts of this chapter are based on the Introduction of the book *Overlooked Cities: Power, Politics and Knowledge Beyond the Urban South*. The book introduction was written by Hanna A. Ruszczyk, Erwin Nugraha, Isolde de Villiers, and Martin Price (Ruszczyk et al. 2021b).

References

Acharya, K.K. (2018) The capacity of local governments in Nepal: From government to governance and governability? *Asia Pacific Journal of Public Administration* 40 (3), 186–197. http://dx.doi.org/10.1080/23276665.2018.1525842

ADRA Nepal. (2018) *Nepal urban resilience project: Scoping study (inception report)*. ADRA Nepal Report. http://adranepal.org/wp-content/uploads/2019/07/NURP-Scoping-Study_Report.pdf (accessed 19 May 2021)

Ali, K.A. & Rieker, M. (2008) Introduction: Urban margins. *Social Text* 26 (2), 1–12. http://dx.doi.org/10.1215/01642472-2007–026.

Apparadurai, A. (2001) Deep democracy: Urban governmentality and the horizons of politics. *Environment and Urbanization* 13 (2), 23–43. http://dx.doi.org/10.1177/095624780101300203.

Asia Foundation (2018) *Diagnostic study of local governance in Federal Nepal 2017*. Kathmandu: Asia Foundation.

Chaudhary, D. (2011) *Tarai/Madhesh of Nepal: An anthropological study*. Kathmandu, Patna Pustak Bhandar.

Datta, S., Jagati, A., Vijaykumar, S., O'Donnell, C. & Basynat, B. for Asia Foundation (2019) *Nepal's transition to federalism: A behavioral approach*. Asia Foundation. https://asiafoundation.org/2019/12/18/nepals-transition-to-federalism-a-behavioral-approach/ (accessed 23 August 2020).

Guneratne, A. (2002) *Many tongues, one people: The Making of Tharu Identity in Nepal*. Ithaca: Cornell University Press.

Hardoy, J.E. & Satterthwaite, D. (1986) *Small and intermediate urban centers: Their role in national and regional development in the third world*. Suffolk: Hodder and Stoughton Educational.

Hardoy, J.E. & Satterthwaite, D. (1989) *Squatter citizen: Life in the urban third world*. London: Earthscan Publication Ltd.

IFAD (2014) Rural poverty in Nepal [Online]. International Fund for Agricultural Development. www.ifad.org/documents/38714170/39972509/Enabling+poor+rural+people+to+overcome+poverty+in+Nepal.pdf/679c83d1-648e-4e67-9a4d-a8430adadd40 (accessed 23 August 2020).

Lawoti, M. & Guneratne, A. (eds) (2013) *Ethnicity inequality and politics in Nepal*. Kathmandu: Himal Books.

Navaro-Yashin, Y. (2012) *The make-believe space: Affective geography in a postwar polity.* Durham, NC: Duke University Press.
Neves Alves, S. & Ruszczyk, H.A. (2021) Theorising from the overlooked city: Generating a research agenda & network on small/secondary cities. *Digital Magazine.* https://93e76a9d-000c-45df-b30c-cce5dbe0e070.filesusr.com/ugd/7cb1c8_ae7c1cea778d4f5ea-c9354008e37e6be.pdf
Nightingale, A., Bhatterai, A., Ojha, H., Sigdel, T. & Rankin, K. (2018) Fragmented public authority and state un/making in the 'new' Republic of Nepal". *Modern Asian Studies* 52 (3), 849–882.
Oldfield, S. (2014) Critical urbanism. In: Parnell, S., & Oldfield, S. (eds.) *The Routledge handbook on cities of the global south.* London: Routledge, pp. 7–8.
Overlooked Cities (2021) Theorising from the Overlooked City Website, overlooked cities.org.
Paterson, S.K., Pelling, M., Nunes, L.H., Moreira, F.A., Guida, K. & Marengo, F.A. (2017) Size does matter: City scale and the asymmetries of climate change adaptation in three coastal towns. *Geoforum* 81, 109–119. http://dx.doi.org/10.1016/j.geoforum.2017.02.014.
Robinson, J. (2006) *Ordinary cities: Between modernity and development.* London: Routledge.
Robinson, J. (2014) New geographies of theorizing the urban: Putting comparison to work for global urban studies. In: Parnell, S., & Oldfield, S. (eds.) *The Routledge handbook on cities of the global south.* New York: Routledge, pp. 57–70.
Ruszczyk, H.A. (2017) *The everyday and events, understanding risk perceptions and resilience in urban Nepal.* Doctoral Thesis. Durham, UK, Durham University. http://etheses.dur.ac.uk/12440/
Ruszczyk, H.A. (2020) Newly urban Nepal. *Urban Geography.* Early view 27 April 2020. https://doi.org/10.1080/02723638.2020.1756683.
Ruszczyk, H.A. Nugraha, E. & de Villiers, I. (eds) (2021a) *Overlooked cities: Power, politics and knowledge beyond the urban south.* Routledge Studies in Urbanism and the City series. Oxon and New York: Routledge.
Ruszczyk, H.A. Nugraha, E., de Villiers, I. & Price, M. (2021b) Introduction. In: Ruszczyk, H.A. Nugraha, E., & de Villiers, I. (eds.) *Overlooked cities: Power, politics and knowledge beyond the urban south.* Routledge Studies in Urbanism and the City series. Oxon and New York: Routledge.
Ruszczyk, H.A. & Price, M.W.H. (2019) Aspirations in grey space: Neighbourhood governance in Nepal and Jordan. *Area* 52 (1), 156–163. https://rgs-ibg.onlinelibrary.wiley.com/doi/full/10.1111/area.12562
Satterthwaite, D. (2019) *Author interview: David Satterthwaite.* www.routledge.com/posts/15986?utm_source=shared_link&utm_medium=post&utm_campaign=B190911461 (accessed 25 September 2019).
Sheppard, E., Leitner, H. & Maringanti, A. (2013) Provincializing global urbanism: A manifesto. *Urban Geography* 34 (7), 893–900. https://doi.org/10.1080/02723638.2013.807977.
Tanaka, M. (2009) From confrontation to collaboration: A decade in the work of the squatters' movement in Nepal. *Environment and Urbanization.* [Online] 21 (1), 143–159. https://doi.org/10.1177/0956247809103011.
Toffin, G. (2013) *From monarchy to republic; essays on changing Nepal.* Kathmandu, Nepal: Vajra Books.
Wikipedia (2020) Bharatpur, Nepal. https://en.wikipedia.org/wiki/Bharatpur,_Nepal (accessed 13 August 2020).

8 When over-researchedness is invisibilised in bibliographic databases

Insights from a case study about the Arctic region

Marine Duc

The last two decades or so have seen the Arctic coming to the fore of public concern and scientific attention. Global change and its economic and political consequences over the region translate into greater interest not only for Arctic resources and seaways but also for the ecological impacts of climate change on its environments and inhabitants' daily life.

Growing academic interest in the Arctic can be observed *in situ*, where fieldwork is conducted. As a researcher involved in the Arctic (Nuuk, Greenland), I have had first-hand experience not only with over-research but also with its impacts in the field: observing it, dealing with it, and considering the ethical and methodological issues it raises, wondering if it can be avoided, and how. Furthermore, it strikes me almost daily, while observing the ever-faster pace at which Arctic research bibliographies grow longer. Large bibliographic databases such as the Web of Science (WoS) provide insights on the growing body of literature dealing with the Arctic. Between 1981 and 2007, Arctic-related publications indexed in WoS were multiplied tenfold, while the total number of indexed publications doubled over the same time period.[1] Ironically, my own research activities are contributing to this growing interest for the region.

First of all, I consider the Arctic to be a heuristic case to reflect on over-researchedness for three main reasons. First, the large number of scientists and publications dealing with the region comes in sharp contrast with the persistent representation of the Arctic as uncharted, despite the constant scientific interest it has aroused since the late nineteenth century. Moreover, scientists have actively participated in the construction of the Arctic as a distinctive place; 'the Great North', simultaneously depicted as the homeland of native peoples which had to be controlled and as a wild and resourceful hinterland (Baldwin et al., 2011; Bocking, 2011; Zeller and Ries, 2014; Cameron, 2015).[2] Paradoxically, considering the large number of research projects dedicated to the region in academic journals and books, the Arctic is still regarded as an empty, remote, and challenging place. This idea permeates literature in different ways. Vocabulary choices immediately spring to mind: when fieldwork is referred to as an 'expedition', while in fact the place studied is Iqaluit, the capital city of Nunavut (Canada), or when the

DOI: 10.4324/9781003099291-9

climate is staged as 'extreme environment'.[3] The way research projects are often linked with the military is another striking characteristic of Arctic research, where sharing facilities with armed forces is quite common. It provides ground for the mutually reinforcing association of science and conquest.[4]

Secondly, the history of science in the Arctic is deeply entangled with colonisation processes and unbalanced research relationships (Krupnik, 2016). Scientific work has been used as a justification to implement assimilation policies even though narratives from neutral and apolitical sciences do persist (Petterson, 2016). Anthropologists, ethnologists, and also glaciologists or botanists have all participated in the construction of the 'colonized Other' as well as the romanticised view of the region (Bravo and Sörlin, 2002; Zeller and Ries, 2014; Cameron, 2015; Andreassen, 2015). Therefore, the Arctic does not escape the categorising gaze of a science production that is structured along with a power relation of race, which makes sense at a global scale (Asad, 1973; Urry, 2003; Andreassen, 2015).

Thirdly, communities perceived as 'hard to reach' are potentially more vulnerable to the negative effects of over-researchedness (Clark, 2008). Except for a few large cities in the Russian North, Arctic communities undoubtedly fall into this category. They can be more affected by the presence of researchers, with increased local impacts of research projects and fieldworks, not only because the colonial context still permeates their daily lives but also due to their remoteness and their small size. Indeed, only a handful of Canadian and Alaskan Arctic towns are home to more than 5000 inhabitants with most settlements counting less than 1000. In Greenland, apart from Nuuk and its population reaching close to 18,000, all towns have less than 10,000 inhabitants, while in fact most of them are under 5000.

Existing literature on the issue of over-research has mainly focused on the local scale, relying on qualitative methodologies, and particularly on ethnography, focusing on places such as refugee camps (Sukarieh and Tannock, 2013; Pascucci, 2017) or neighbourhoods and urban communities (Neal et al., 2016; Desvaux, 2019). I argue for placing the analysis at another scale, and observe how visible (or invisible) over-researched places are when the issue is examined from the other end of the research process: published work, considered as the most visible part of the research process (Latour and Wooglar, 2006). As large bibliographic databases now index most publications, they provide us with an interesting source to conduct such inquiry. So far, they have mostly been used to study academic and institutional networks in a geography as science production perspective (Matthiessen et al., 2002, 2010; Eckert et al., 2014; Beauguitte and Maisonobe, 2017). Here, I will focus on a different set of information offered by these databases—the keywords that refer to fieldwork places and spaces. Hence, I consider bibliographic databases as a vast library where various traces of fieldwork remain once data collection is complete. To map over-research at the macroregional scale, I have developed a quantitative exploratory method based on scraping[5] data from WoS and then built a toponymic database to finally process this data.

For mapping and analysing the geography of Arctic fieldworks as they appear in the WoS, I rely on both critical data studies and critical toponymy. Critical data studies aim at challenging the idea of Big Data as a neutral set of information and 'track the ways in which data are generated, curated, and how they permeate and exert power' (Iliadis and Russo, 2016: 2). Critical toponymy 'emphasizes the spatial politics of naming and the social production of place' (Rose-Redwood, 2011: 34). It seeks to excavate the issues not only of territorial appropriation but also of contestation or resistance which are expressed in place names, as well as the processes leading to the choice of a particular name, for a specific place. In the context of a developing Geoweb, the way these place names circulate also expresses power relations. These are explored by Critical toponymy (Noucher, 2020). I suggest that large bibliographic databases contribute to the invisibilisation of over-research in specific places. As circulation devices, they regulate which place names are visible (language and scale of reference) and play an active role in legitimising some words over others to designate and qualify researchers' fieldwork. In this chapter, I will show that exploring the geography of science through the place names mentioned among the WoS' keywords reveals how the normative effects of publication injunctions impact the issue of over-research.

The goal of this chapter is twofold. Not only does it explain and demonstrate how large bibliographic databases can be of use to objectify the feeling of over-researchedness expressed in a number of Arctic places but it also presents a methodological proposal to investigate the way in which the indexing practices of big bibliographic databases can spatially structure research activities.

The first part of the chapter considers how critical data studies can be instrumental in addressing over-researchedness through an examination of the toponyms found in the keywords of indexed publications. The second part focuses on the dual level at which our results can be read, moving from a descriptive to an interpretive reading. The final discussion highlights the methodological issues raised by the study and suggests how a change in indexation practices could help 'rescale' the way in which fieldwork emerges in keywords and therefore make over-researchedness more visible.

Place researchedness and critical data studies: bibliographic databases as regulators of place name flow

Over-researchedness and the Arctic

While many studies, especially those based on a qualitative approach, have developed reflexive works about the complex relationships between researched people and researchers, very few have explicitly addressed the issue any further than the local scale. Existing literature has mainly been concerned by the factors which caused the communities to complain of being over-researched (Clark, 2008; Sukarieh and Tannock, 2013; Neal et al., 2016). They have raised the issue of over-researchedness from two complementary perspectives. The first one focuses on the consequences observed on researched communities and questions the impact of research work conducted in specific places. It underlines the risk of 'research fatigue', when 'individuals and groups become tired of engaging with research and it can be identified by a demonstration of reluctance towards continuing engagement with an existing project, or a refusal to engage with any further research' (Clark, 2008: 956). Such fatigue is attributed not only to previous involvement with projects dealing with similar issues, repeated lack of outcomes or benefits, misrepresentation of community members but also to practical causes such as time-consuming participation or guidance fatigue (Smith, 2008; Sharp and Murdoch, 2006; Clark, 2008; Sukarieh and Tannock, 2013). The second perspective examines the production of knowledge itself, addressing power relations, research ethics, and adapted methodologies to avoid research fatigue (Sukarieh and Tannock, 2013; Aiken, 2017). This chapter follows this line of discussion, addressing particularly the scientific production of geography in the Arctic, as it appears in the Web of Science.

Sukarieh and Tannock (2013) identify three characteristics causing some communities to be more at risk than others, of being negatively impacted when many research projects are conducted in their area, in particular being over-researched. These characteristics are poverty and marginalisation, past crisis and/or resistance or adaptation to it, and accessibility to outsiders. Most Arctic regions present two or three of these characteristics. Poverty and marginalisation are significantly more prevalent in comparison with the national standards of the country, especially in Indigenous communities all over the Arctic (Poppel, 2015; Larsen and Fondahl, 2015). Colonisation processes along with their psychological, economic, and social consequences undoubtedly form the ongoing crises rooted in the past and currently increased by global changes that are operating with more intensity in the Arctic than anywhere else (IPCC, 2018). As for accessibility to outside researchers, it has an ambiguous effect in the Arctic where isolation and accessibility costs are appealing factors because they nurture a sense of curiosity for the region. Doing fieldwork in the Arctic can also reactivate the exoticising 'explorator's culture' (*la culture de l'explorateur*: Lefort, 2012), according to which the researcher embodies the original discovery and reactivates the disciplinary anthropological myth of the distance of the subject and that of otherness, as a guarantee of the quality of research (Fabian, 1983; Sontag, 1994; Geertz, 2003).

Local institutions and researchers have highlighted how research fatigue can become a reality in many Arctic communities not only due to the significant number of ongoing projects but also due to the way knowledge production itself reproduces power relations at play. The latest *National Inuit Strategy for Research* (Inuit Tapiriit Kanatami, 2018) states that 'Inuit in Canada are among the most researched people on the Earth', while 'the primary beneficiaries of Inuit Nunangat research continue to be researchers themselves' (2018: 5).[6] As in other parts of the world, research is also increasingly becoming 'a dirty word' (Smith, 2008) as 'it is inextricably linked to European Imperialism and Colonialism' (1999: 1). Indigenous communities have been 'researched to death' (Goodman et al., 2018) and several Indigenous scholars have also pointed out a tendency to prefer 'damage-centered' research projects (Tuck, 2009). While these may stem from good intentions, they can also reproduce pitiful representations of Indigenous communities (see e.g. Louis, 2007; Wilson, 2008; Tuck, 2009; Smith, 2008; Tuck and Yang, 2012; Asselin and Basile, 2018). These scholars are calling for more participative and community-based research practices carried out with increased awareness of the political and social consequences of the research process. Yet, participatory research is not a magic shield against research fatigue and unbalanced power relations involved in any research process (Titterton and Smart, 2008; De Leeuw et al., 2012).

However, an ongoing change can be observed in the research practices led by Indigenous communities themselves along with Indigenous, decolonial, and feminist scholars to critically discuss the way Indigenous peoples are positioned 'within research and (re)produced and represented as research subjects' (De Leeuw et al., 2012: 181) but much remains to be done. A recent study shows that young researchers conducting fieldwork in Alaska and Northern Canada tend to have more experience and knowledge about working with Indigenous communities than those working in Scandinavia and Northwestern Russia. In addition, social science researchers are more informed about and make greater use of participatory and community-based research. The authors suggest that these unbalanced results are due to the different legal frameworks and ethical guidelines. In addition, most researchers in the Arctic are still outsiders, *Qallunaat* (i.e. non-Inuit) in the Inuit language (see e.g. Nungak, 2006; Graugaard, 2020), a factor that may increase research fatigue. Indeed, outsiders are more often exposed to the need for practical services (translation and guidance). Their lack of local involvement—or a limited one—can also impact the relevance of their research topic and raise difficulties for them to adjust to local rhythms or social codes. Moreover, being a *Qallunaat* exposes the researcher to perpetuating a dominant position of power in the conduct of fieldwork. Carrying out ethnography research or interviews, thus represents a specific area where power relations such as gender or race (among others) are (re) made and where social positionalities are performed (Best, 2003; Fenstermaker and West, 2002).

In addition to changing research practices, another answer to over-researchedness is the regulation of research through research licensing and raising awareness among scientists about the ethical issues inherent to research, especially in small Arctic communities. This is strongly implemented in Northern Canadian territories. In Nunavut and in the Northwest Territories, any research project must be granted a license by Inuit local and regional authorities and approved by an academic ethics committee. In contrast, this is not the case neither in Northern Quebec and Labrador nor in Yukon and in the USA (Alaska) (Collignon, 2010). In Greenland, only projects related to health issues require to be approved by an ethics committee. Ilisimatusarfik, the University of Greenland, offers workshops for PhD students, mainly in Danish and open to all PhD students conducting research in Greenland. Some directly target ethical issues and unbalanced research relations. In 2018, PhD courses such as 'Research Cohort, Reproduction, and Environmental Health', 'Community Involvement, and Population Trends Across the Arctic', and 'Multidisciplinary Research' were organised, addressing ethical challenges for researchers: 'This session will provide a brief overview of participatory research methods that foster respectful community engagement and reciprocal learning, including diverse Arctic project examples and opportunity for active case study discussion and consideration of potential ethical challenges' (Ilisimatusarfik, Introduction to PhD course II, 2018).[7] In 2013, the parliament of Greenland voted to establish a research council, 'to ensure a closer dialogue between the research and the Greenlandic public' and 'ensuring the best possible knowledge base for the development of the Greenlandic society'.[8] All these committees aim at regulating the unbalanced power relations at play in research relationships. They represent one mechanism to try to slow down the 'extractive' (Smith, 2008) practices of researchers in the Arctic. However, although they attempt to prevent the recurrence of past abuses (absence of consent, objectification, and dehumanisation), especially in health studies historically connected to physical and racial anthropology (Andreassen, 2015), they do not set any quantitative limitation to the number of research projects that can be conducted in one location at a given time.

Place names and over-researchedness: the politics of bibliographic databases

Over-researchedness implies the idea of measuring a study based on 'quantity' of conducted research. Large bibliographic bases appear as an interesting source to deepen the measure of research as they present a set of data that expands over a long period of time (WoS data goes back to 1900) and includes a wide range of information, allowing for comparison. Bibliographic databases carry various kinds of geographical information: authors' institutions (often with their address) and researched places mentioned in full texts, abstracts, and keywords. For the purpose of this study, I focus on the latter. Several works have already used this information to provide insights on the landscape of research in polar regions. They signal the faster growth of indexed publications related to polar regions between 1981 and 2007, which have increased fourfold while the total number of publications in WoS only doubled (Aksnes and Hessen, 2009). The pace is even faster when it comes to the Arctic, with related publications increasing tenfold over the same period of time. Other works looked into institutional affiliations to explore the geography of science production at a global level (Matthiessen et al., 2002, 2010; Eckert et al., 2014; Beauguitte and Maisonobe, 2017), and in particular, the research networks north of the Arctic Circle (Beauguitte and Maisonobe, 2017). However, specific places and place names related to empirical fieldwork have been rather overlooked so far. This is why I suggest examination of place name circulation in the WoS to explore what they reveal about the presence of researchers in Arctic fieldworks. Focusing on keywords (instead of abstracts or titles) can thus account for visibility issues in indexing practices. Keywords refer in fact to the choices made by researchers not only to locate their work but also to be visible and easily identifiable upon bibliographical research. Choosing keywords follows a logic of classification based on the language and the grapheme used, the scale of reference chosen, and the fields of research named. This is not a neutral process.

Focusing on power relations involved in the place naming processes, critical toponymy uncovers the antagonistic interests involved in the choice and use of place names (Berg and Vuolteenaho, 2009; Rose-Redwood et al., 2010; Rose-Redwood and Alderman, 2011; Giraut and Houssay-Holzschuch, 2016). Matthieu Noucher (2020) expands the analytical tools of critical toponymy to examine the way place names circulate, on the web in particular. Using a Foucauldian framework, he stresses how, in the digital age, cartography is reconfigured as a power technology. Navigating systems, geolocation apps, or online databases such as GeoNames (which gathers more than 25 million place names) contribute to normalizing the use of specific place names which then become references. Considering the 'place name flow' or the way toponyms spread and circulate in the Geoweb, Matthieu Noucher suggests that online nomenclatures participate in normalising the online landscape of toponymy by associating a place with a specific name and invisibilising alternative place names. Therefore, they shape our perception of the world (Goodchild and Hill, 2008).

Scholars themselves take part in this place name flow. First, when they research place names directly for the purpose of their research. For example, Béatrice Collignon was involved in collecting Inuinnait toponyms in Northern Canada as part of her PhD project. She worked with local Inuit communities to have their place names officially recognised and eventually printed on Canada's maps (Collignon, 2004). But scholars are also involved in conflicts surrounding place names when they disseminate the results of their studies and index their publications. The issue of naming Greenland presents a striking illustration of this. Inspired by postcolonial and/or decolonial work, a growing number of papers and books opt for the Kalaallisut name of Greenland, *Kalaallit Nunaat* (meaning the land of the Kalaallit) rather than 'Greenland', the English translation of the Norse-Danish colonial name, *Grønland* (see for instance. Grydehøj et al., 2020). The term 'Kalaallit Nunaat' serves either as a performative tool of territorial reappropriation or as a way to recognize local territorial reappropriation. However, this falls short of being the perfect decolonial move. The ethnonym 'Kalaallit' (sing. *kalaaleq*) refers to only one of the three groups living in Greenland (Kalaallit on the West Coast, Inughuit in North and the Thule Region, and Tunumiit on the East coast). Therefore, the fact of calling Greenland 'Kalaallit Nunaat' participates in invisibilising and marginalizing the two other historically and linguistically distinct groups. Moreover, the spread of this name is deeply entangled with colonial presence. Indeed, it is generally admitted that the name *Kalâdlit Nunæt* was suggested by a missionary at the end of the eighteenth century, after he had noticed that Greenlanders did not have any other name than *Nunarput* (our country) to designate their own land (Kleivan, 1977). The West Coast influence was central in spreading the name because it concentrated most of the population, the press, and the catechists who were educated at Nuuk and Ilullissat Seminars[9] (Kleivan, 1996; Thomsen, 1998). For Inge Kleivan, 'the name Kalaallit Nunaat, "land of the kalaallit" is an example of cultural imperialism among the Greenlanders themselves' (Kleivan, 1977: 199).

The choice of place names conveys the researchers' relationship with their fieldwork: the keywords they choose to index their publications are not disconnected from their fieldwork experiences. Hence, the hypothesis that large bibliographic databases also shape the norms of place name flow. The mechanism is the same as the one described by Matthieu Noucher (2020). Although these databases are not a place name database as GeoNames can be, they too make certain place names more visible than others. Just as GeoNames does, they operate as regulating devices of place name flow as they are the bottleneck of the place name choices made by the authors to ensure the visibility of their research towards their peers.

From descriptive to interpretative reading of science databases: exploring the blank of databases

The idea behind this was to conceive a method to analyse research conducted in the Arctic and identify the most researched places. To explore the place name flow in large bibliographic databases, I decided to focus on WoS, one of the most comprehensive science databases, along with Scopus. While Scopus covers a larger number of journals, WoS was for many years the only citation and publication database, and therefore, spans over a larger period of time for all domains of science, since 1900 (Aghaei Chadegani et al., 2013). However, WoS is well known for its linguistic and geographical bias (Maisonobe, 2015), and so is Scopus. Unlike other databases such as the European Reference Index, both mainly cover English-speaking journals and journals published in the United States and in the United Kingdom (Dassa et al., 2010).

The data extracted can be analysed at two different levels. At the descriptive level, it pinpoints the places and themes explored by research and draws the geography of science production that stems from fieldwork. At the interpretative level, it addresses the normative effects of place naming in scientific publications related to the Arctic.

Methodology: researching researchers' practices through bibliometrics

Our corpus expands from 1918 to 2018 and gathers over 55,000 publications that match the request 'Arctic' in the WoS search engine on keywords associated with any type of document available in the Core Collection (articles, proceeding papers, book chapters, book reviews, research notes, meetings minutes and abstracts, etc.). Several collections structure the WoS ('all databases', 'Core collection', 'KCI-Korean Journal Database', 'Medline', 'Russian Science Citation Index', and Scielo Citation Index'), each with a different focus (disciplinary or geographic). Each collection is also based on different indexes, measuring the citation of each publication. Although the All Databases Collection is the largest, I decided to work with the Core Collection because it provides more accurate and exhaustive metadata. For each paper, all the authors and their affiliations, all abstracts, all keywords (when available), all acknowledgements (grants and funding agencies), and all cited references are included.[10] Since the corpus is built on the self-identification and indexation practices of the authors themselves, I have no influence on how the geographical area shaped by the corpus is delineated. However, relying on one keyword—Arctic—to select the documents that would be included in the corpus means that published works dealing with the Arctic but not referenced with this keyword would not be included.

As part of this bibliographic 'Arctic Corpus', I collected all the 'author keywords' from all scientific productions indexed between 1918 and 2018 through scraping in R language. Scraping did not retain the documents without any keywords (50.3% of the corpus). The focus on keywords presents the advantage of revealing the normative effects of the limited number of terms often available to reference one's field of investigation. As they are more precise than titles and less constraining to use than abstracts, keywords also carry the visibility issues of fieldwork at stake in large databases. This leads us to the last step. I conducted an inventory of the place names likely to relate to a researched place within the keywords database. The keywords were organised through a textual analysis of the corpus using the Iramuteq Open Source Software based on R language. As it categorizes word types according to their form (adjectives, adverbs, pronouns, nouns, etc.), most place names fall under the 'unknown forms' category, together with vegetal or animal species, technical terms, and geological areas. Yet, some of them are identified as 'nouns'. The next step of the sorting task thus consisted of creating a second corpus in which keywords from the 'unknown form' and the 'noun' categories were put together. I then proceeded manually to create an inventory of toponyms that are likely to relate to researched places. It includes all the non-generic place names of all scales that designate a segment of space (sites, regions, places, and country names). Every place name that could indicate a research device on the field, including marine areas, was also kept. Consequently, a number of place names were not included: those referring to past geological conformations ('Gondwana') and adjectival forms ('Swedish'). I proceeded on a case-by-case basis, checking whether or not the term could be related to a place located in the Arctic (e.g. 'Rayleigh' was not selected as it refers to either a British town, physical properties, law, or a measurement unit named after a British Physicist). Unlike the previous ones, this last step of data selection was clearly influenced by my own positionality and my own definition of fieldwork.

Mapping researched places and themes addressed by Arctic scientists

The first results show how bibliometrics can serve as a tool to objectify and measure research interest for a specific region, both in terms of disciplines and geographic areas. Figure 8.1 shows how, in the corpus gathered, Geoscience categories[11] largely dominate Arctic research. Yet, this should be read with caution as the WoS has a well-known tendency to over-represent natural sciences. Moreover, its use of the analysis of social sciences and humanities production trends has been criticised (Archambault and Larivière, 2010).[12] Moreover, its linguistic bias in favour of English has increased over the years (Maisonobe et al., 2016; Maisonobe, 2015). It is therefore important to remember at this point that WoS only indexes a selected portion of global science production (Maisonobe, 2015).

Figure 8.1 also shows a clear increase in the number of Arctic-related publications, which started in the late 1980s. The comparison with the full range of

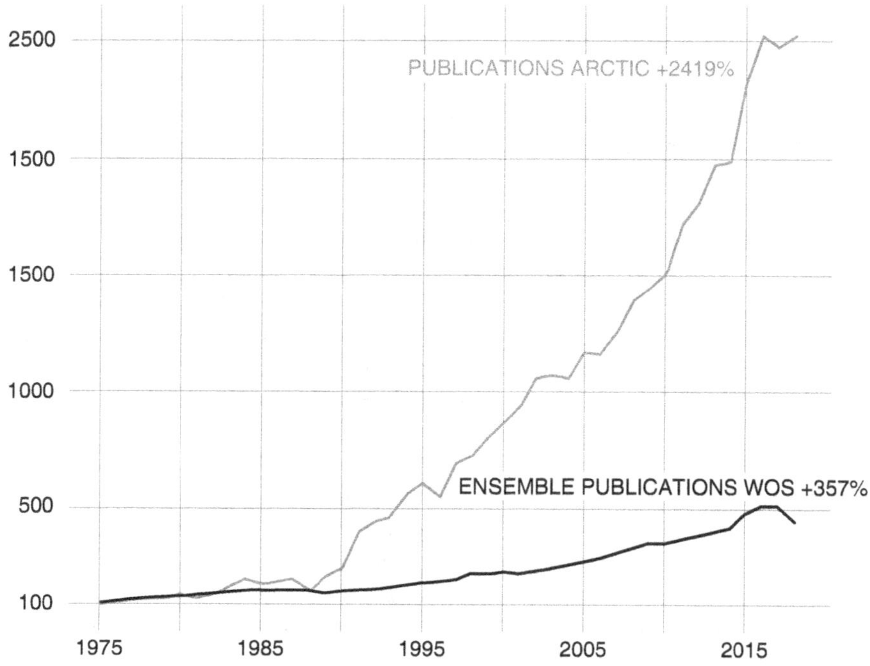

Figure 8.1 Indexed Arctic-related publications in the Web of Science Core Collection: over-representation of geosciences (1) and breakthrough of Arctic-related publications (2).

Source: ©M. Duc and G. Le Campion 2020

indexed publications within the Web of Science over the same period of time is striking. Interestingly, this breakthrough of Arctic-related publications follows the time frame of growing environmental awareness and its institutionalisation at the global scale. In 1977, the Inuit Circumpolar Council was created in reaction to the growing presence of oil extracting companies in the North American Arctic (Petersen, 1984; Dahl, 1988; Morin, 2001). A few years later, in 1983, the United Nations funded the World Commission on Environment and Development, which resulted in the Brundtland Report in 1987, and the implementation of the Intergovernmental Panel on Climate Change in 1988.

Turning now to the keywords of the second corpus, focusing on place names, we can see whether this over-representation of geosciences in Arctic research translates in the field, in researched locations, and how. Are researchers from social sciences and humanities investigating the same places biologists or meteorologists do? In fact, as it can be seen in Figure 8.3, very few differences are noteworthy between the two groups. Even more surprising, most-researched places in the Arctic appear to be located not on the land but instead within maritime areas (see Figures 8.2 and 8.3). This could be the result of the socio-economic

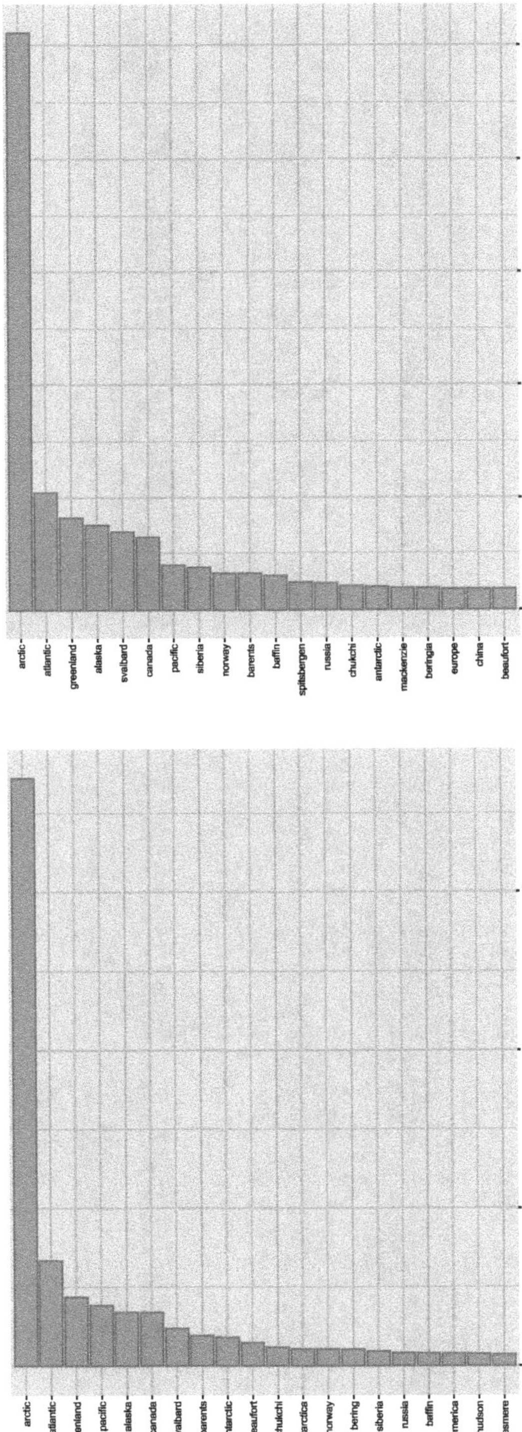

Figure 8.2 Chart of toponyms in the keyword database per number of mentions for natural sciences (left) and social sciences and humanities (right).

Source: ©M. Duc and G. Le Campion 2019

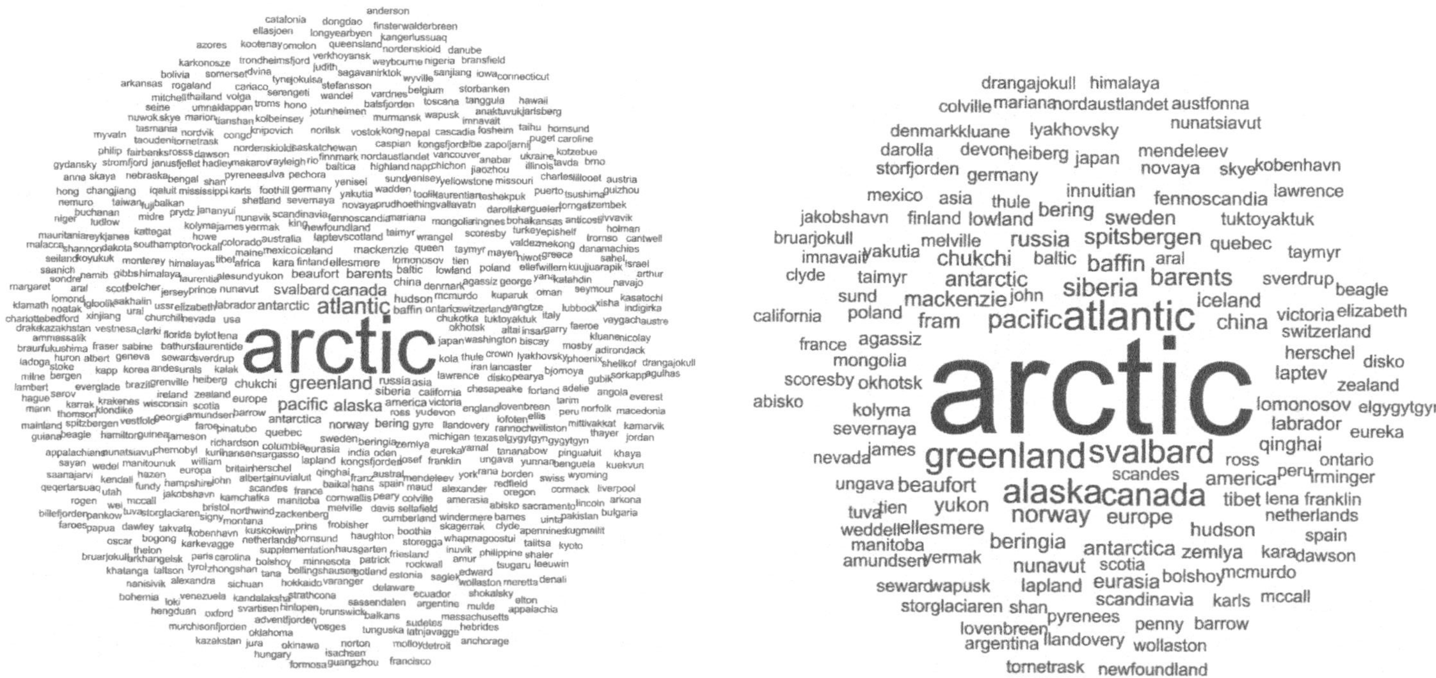

Figure 8.3 Papers' keyword word clouds for natural sciences (left) and social sciences and humanities (right).

Source: ©M. Duc and G. Le Campion 2019

and cultural relevance of the sea and sea-ice for Arctic societies and of the large surface covered by the sea around the geographic North Pole. But there is more to it as the interpretative approach of the corpus will show. In Figure 8.3 a few keywords relating to place names outside of the Arctic also appear. This is probably an artefact of comparative research projects involving other parts of the world, mainly 'Antarctic', 'Switzerland', and 'Germany'.

Altogether, the result of this approach of Arctic research through a descriptive analysis of keywords scraped from a large bibliographic database offers a regionalised state of the art. Most popular scientific domains are identified and so are most studied regions. And yet, precise fieldwork locations do not appear. Therefore, one may wonder what this regionalised picture of Arctic science says about over-researchedness. Oceans are over-represented among researched places in our database, which does not provide any ground to understand research fatigue, lack of benefits for inhabitants, or misrepresentation of community members. How can such databases help identify operational parameters to set a threshold beyond which a place can be considered 'over-researched'? Should human presence be the only relevant factor to detect over-researchedness? Should the material impacts of some research projects, such as the depletion of an archeological site, the loss of a geological formation due to over-frequentation, pollution of soils and rivers caused by research activities, and sometimes the mere presence of a large camp over several summers, be taken in account?

In this light, the corpus built as a tool to measure and research over-researchedness in the Arctic appears to be inadequate. But why is that so? This brings us to take a new look at large bibliographic databases and consider them as a device which organises spatially and orders scientific information. Therefore, the issue that needs to be addressed is the politics of research dissemination. How do indexation practices and the logics of keyword selection translate into a scalar gap where over-researchedness becomes invisible?

The black holes of databases: keyword choices, over-researchedness scrambling, and the ordering of science

In the corpus gathered, I have not found a single Indigenous place name. This is highly significant and leads to a surprising number of absentees among place names. Well-known research sites such as research stations, regional capitals, and emblematic Indigenous settlements do not have a substantial presence in the keyword database. For example, 'Thule' only appears 25 times the same as 'Tibet'. The town of Qaanaaq, close to Thule, which is a popular research base does not appear once in the corpus, although it hosts the Arctic Hydrology Research site of the University of Copenhagen and the Geophysical Observatory managed by the Danish Meteorological Institute. The main cities of the North American Arctic also have very few occurrences in the database: Iqaluit and Inuvik (Canada) only appear four times and Nuuk (or Godthåb, former colonial name of Nuuk, Greenland) never does. Although, in at least the last decade, many social science conferences have highlighted urbanisation as a major trend in the twenty-first century

Arctic and therefore represents a key research topic for the years to come, the keyword itself, 'urbanisation', appears only three times in the corpus. A twofold process of thematic (urbanisation or mobility for example) and spatial (specific places) invisibilisation seems to be at work here. More precisely, there seems to be a scale gap between the scale of fieldwork and the scale of paper referencing. Keyword ranking shows an obvious preference for the regional scale when it comes to referencing one's paper and the subsequent erasure of fieldwork sites in bibliographic databases. Whether this scalar gap results from the way the database itself is set up, from the analytical model, or from the referencing practices of research publications remains to be addressed. Further research is needed on this key point.

However, the disappearance of the microscale reveals the visibility game at play when researchers select their keywords along the publication process. Keywords are meant to encapsulate the complexity of a research result in a few words and to ensure their maximum visibility. Publishers and institutions tend to advise authors to choose keywords that will promote the visibility of their work, a guideline that promotes the use of terms already referenced, rather than uncommon ones. A publishing guide edited by the Royal Geographical Society is going in that direction (Blunt et al., 2015) and keyword recommendations on Wiley's Website (2020) as well:

> your keywords should employ natural language and blend seamlessly into your article . . . Great keywords capture the attention of search engines while also accurately conveying the content of your article the correct keywords may give your article better circulation among internet searches, which, in turn, could elevate your expertise in your chosen subject.[13]

Keyword uses are based on the logic of cumulative visibility. In a nutshell, the more an article is cited, the stronger is its impact factor and the more it becomes visible in large databases and search engines, the more it gets consulted. Consequently, the scalar scramble between empirical data collection and production referencing enables us to understand how researchers who work in the Arctic ensure that their publications will be easily visible, although their work takes place in a region that remains pictured as uncharted. But in fact, this scrambling ultimately operates like a self-fulfilling prophecy. As they sense the lack of awareness of the toponym referring to the place where their empirical survey has taken place, researchers decide to reframe it at a smaller scale when it comes to referencing their published results. Such a practice is instrumental in constructing the Arctic as a smooth and homogeneous surface in bibliographic databases. It also makes the quantitative measure of over-researchedness almost impossible to perform due to the lack of the visibility of specific places where research actually takes place.

This can be labelled 'phantom geographies of fieldwork and researched sites'. These phantom geographies clearly reveal the extent to which large bibliographic databases take part in ordering scientific production through the normalisation mechanisms of keyword circulation in the numerical landscapes of publication, from publishers' recommendations to search engines. The idea is not to regard big bibliographic databases as the sole cause of the invisibility of fieldwork in their collections but rather as a device that crystallizes a larger set of publishing practices. The performative effect of measuring instruments, which puts the social world in order (Desrosières, 1993) has been described for other devices such as census statistics (Simon, 2003) or for information circulating in place name databases on the Geoweb (Noucher, 2020). While digital technology is developing, this world ordering is accelerating, making it 'measurable in everything' (Cardon, 2015, quoted by Noucher et al., 2019). Our study demonstrates that this encoding effect is also involved in the indexing processes of publications, in large bibliographic databases. In this case, it is not a social category (such as racial or gender categories in census statistics) that operates as an ordering device but the reference scale that becomes a system of meaning in indexing, a spatialised categorisation mechanism legitimizing certain toponyms over others.

Discussion: ordering science production and making fieldwork visible

Deepening data collection through mixed methods

The findings of this exploratory project would benefit from more thorough research, particularly with regard to its methodological framework. It appears that turning to bibliographic databases to objectify research practices needs to be done by examining at least two levels of numerical representation. On the one hand, through the Web of Science's technical indexing system, and on the other hand, through the data collection system itself, which carries its own categorisation effects. Three main methodological problems have been emphasised throughout this chapter. First, data scraping was based on the co-use of the 'Arctic' keyword, which has a strong selection effect. Secondly, the IraMuTeq software cannot handle compound words, which means that a number of keywords have been left out from the second toponym database. And thirdly, our own, hence specific, definition of what fieldwork has also affected toponyms selection.

Further research shall aim at improving data collection devices. Given the large period of time covered by the WoS Core Collection, a chronological perspective on referencing should also be developed. Looking at how referencing practices have changed throughout the years might reveal interesting temporalities in the growing visibility of specific keywords. Also, turning to qualitative methods to supplement the findings drawn from the quantitative approach of such databases will most certainly prove to be very fruitful. For example, from a sociology of science perspective, it would be quite interesting to investigate the importance given by authors to their choice of keywords to explore their referencing and publishing practices in greater detail as well as investigating the way publishers and institutions use large bibliographic databases in the process of defining strategies and policies.

Contesting the invisibility of over-researchedness

While bibliometric indicators have become increasingly important over the last 40 years or so, in the context of internationalisation and neoliberalisation of research (King, 2009; Jarvis, 2014; Harari-Kermadec, 2019), the problem is not the scalar gap itself but rather the fact that it is potentially instrumental in ranking, evaluating, and even channelling research on selected topics through funding choices. This scalar gap may have disruptive effects on the way in which hot research topics are signposted.

As an illustration, let us examine the *International Opportunities for Danish Arctic Researchers. Innovation Centre Denmark: Mapping Arctic Research Activities in Five Countries* (2019) report. It is based on an analysis of the Scopus database and looks at the institutional affiliations of Arctic scientists, along with the keywords that identify researchers' thematic and geographical fields of specialisation and linking this information to the researcher's nationality. Without considering any of the biases of the bibliographic database, the report aims at measuring the Arctic research activity conducted by several countries in order to assess the value of cooperation. The interest is then established according to not only the existing collaborations, the history of involvement in the region, and the available funding and infrastructure but also bibliometric indications that point to the thematic areas of research involvement. The report then concludes that the different maritime passages and oceans but also 'Greenland' and 'Alaska' are considered as 'big', 'main', 'central topic', or 'big thematic area'. In this report, the most interesting countries to establish partnerships with are the ones that produce a large number of publications referenced in Scopus and co-publications with Danish researchers. Geographic themes/areas are instrumental in supporting the process by identifying the areas that projects should focus on. This example shows how geographic information held in large bibliographic databases can be used to organise research agendas and therefore standardise its value.

In this age where research policymakers increasingly rely on bibliometrics to measure scientific production and target research investments on specific topics, questions can be raised about the consequences of the scalar jump of keywords referencing in these large databases. If statistical instruments, indicators, and other accounting devices become a governmental technique under the guise of efficiency and improved management (Lascoumes and Le Galès, 2004), what could be the effects of measurement on the spatial distribution of research? Could this not lead to identifying research topics without even considering the risk of over-researching particular places? Or to a legitimisation of the regional scale, in the conduct of surveys? Or even, as could be the case with the aforementioned report to reinforce the trend to research, over and over, themes and fields already thoroughly studied?

Consequently, I call for a change in our referencing practices as a way to oppose the current invisibilisation occurring in bibliographic place name flow. It is difficult to control the various uses of large bibliographic databases; yet, it would be rather easy to control the geographic information they will convey once we, researchers, have chosen our keywords. Although I am hereby challenging the common keyword choice recommendation meant to ensure that our papers get optimum visibility in search engines and databases, I would like to promote the use of a precise designation, at a micro-scale level, regarding the places where fieldwork has been conducted, along with including Indigenous toponyms, whenever possible. This would contribute to preventing researchers from having the upper hand over their fields of investigation and over those who live in and around them. In addition, this could change the way the act of selecting keywords is usually considered. Instead of a rather useless burden for researchers or a move to support their career strategy and make themselves visible through research, it could become an opportunity for them to reflect upon the aspects of their work that they wish to bring to the fore.

Acknowledgements

This chapter stems from a paper presented at the RGS/IBG 2018 Annual Congress, along with Professor Béatrice Collignon (University of Bordeaux Montaigne, Passages Joint Research Unit, UMR 5319). This paper also forms the basis of 'À la recherche de la sur-enquête dans les grandes bases de données bibliométriques. Propositions exploratoires à partir d'un cas d'étude sur l'Arctique", Annales de Géographie *2021, 742 which is more methodologically oriented.*

Figures 8-1, 8-2 and 8-3 were developed by Grégoire Le Campion, CNRS Research Engineer at Passages Joint Research Unit.

Notes

1 Data from Web of Science, 2018. In 2016, the agreement between the National Centre for Scientific Research (CNRS) and Thomson-Reuters was renewed, giving French researchers access to the Web of Science Bibliometric Database for four more years (February 2016–February 2020) (https://negonat.inist.fr/spip.php?article90).

2 Such ideas are also disseminated to the general public: throughout emergent tourism activities (Watson et al., 2002), emblematic landscapes of climate change (Stoddart and Smith, 2016) or more broadly, through industrialization and extractive activities (Nuttall, 2010), to name but a few.

3 See, for example, this debriefing note about the fieldwork of a PhD student from UQAM University in Iqaluit: www.actualites.uqam.ca/2019/expedition-scientifique-grand-nord-canadien.

4 For example: the Villum Research Station, a research facility at the military outpost Station North at the tip of Northeastern Greenland, operated by Aarhus University and the Danish Defense/Arctic Command (https://villumresearchstation.dk/).

5 Web-scraping is an extracting method to harvest data from a webpage using a code. Marketing and consumption studies mostly collect spatialised data. They rely increasingly on scraping methods (Mermet, 2017).

6 The report also uses bibliometry to provide evidence of a growing volume of research conducted over time. It highlights that 'the number of peer reviewed publications and dissertations that focus on Inuit and Inuit Nunangat has increased at a rate higher than the increasing population of Inuit' (Inuit Tapiriit Kanatami, 2018: 17). According to the report, there was one publication or dissertation for seven Inuit in 1996; in 2011, it had increased from one to three. Two main types of consequences are highlighted. The first one is related to knowledge production (lack of free, prior, and informed consent, lack of control, misrepresentation of communities, or poorly adapted solutions, and unbalanced access to research outputs, data, and information). The second one refers to the material (lack of) effects of research, such as unbalanced access to funding, persistence of inequalities, and poverty for Inuit versus career advancement for researchers.

7 https://uk.uni.gl/events/2018-hosted/june-hosted/phd-course-ii.aspx.

8 Quote from The Greenlandic Research Concil's website, last consulted 2018: https://nis.gl/en/

9 Some of these catechists were of mixed origin and were born in Greenland (Petterson, 2014). This simple example signals how the relation between the categories of the colonizer and the colonized in Greenland is far more complex than one of mere opposition.
10 Extra metadata can be added as well: ORCID identifiers, additional funding data from MEDLINE and Researchfish, and unified institutions' names. See https://clarivate.libguides.com/webofscienceplatform/woscc for a full picture.
11 In each Index, WoS classifies publications according to disciplines. For the purpose of this study, I used WoS disciplinary categories.
12 Archambault and Larrivière stress some peculiarities of publication patterns in humanities and social sciences: the importance of books and series in knowledge dissemination at a longer post-publication rate, the local relevance of social sciences and humanities knowledge, which translates into a trend among these scholars to publish more in their mother tongue and/or in journals with a limited distribution, than the general average.
13 Wiley's website is available here: www.wiley.com/network/researchers/preparing-your-article/how-to-choose-effective-keywords-for-your-article (2020).

References

Aghaei Chadegani, A., Salehi, H., Yunus, M., Farhadi, H., Fooladi., M., Farhadi, M. and Ale Ebrahim, N. (2013) A comparison between two main academic literature collections: Web of Science and Scopus databases. *Asian Social Science*, vol. 9, no. 5, 18–26.

Aiken, G. T. (2017) Social innovation an participatory action research: A way to research community? *European Public & Social Innovation Review*, vol. 2, no. 1, 17–33.

Aksnes, D. W. and Hessen, D. O. (2009) The structure and development of polar research (1981–2007): A publication-based approach. *Arctic, Antarctic, and Alpine Research*, vol. 41, no. 2, 155–163.

Andreassen, R. (2015) *Human Exhibitions: Race, Gender and Sexuality in Ethnic Displays*, Farnham: Ashgate.

Archambault, E. and Larivière, V. (2010) The limits of bibliometrics for the analysis of the social sciences and humanities literature. *World Social Science Report*, 251–254.

Asad, T. (1973) *Anthropology and the Colonial Encounter*. London: Ithaca Press.

Asselin, H., and Basile, S. (2018), Concrete ways to decolonize research, *ACME: An International Journal for Critical Geographies*, vol. 17, no. 3, 643–650.

Baldwin, A., Cameron, L. and Kobayashi, A. (eds.) (2011) *Rethinking the Great White North: Race, Nature, and the Historical Geographies of Whiteness in Canada*. Vancouver: UBC Press.

Beauguitte, L. and Maisonobe, M. (2017) Les villes arctiques dans la production scientifique mondiale. Une périphérie en voie d'intégration? Paper presented in Pôles Urbains Project seminar (ANR PUR n° ANR-15-CE22–0006, 2016–2019, PI: Y. Vaguet).

Berg, L. D. and Vuolteenaho, J. (eds.) (2009) *Critical Toponymies: The Contested Politics of Place Naming*. Farnham: Ashgate Publishing.

Bergsland, K. (1986) *Comparative Eskimo-Aleut Phonology and Lexicon*. Helsinki: Suomalais-Ugrilainen Seura.

Best, A. L. (2003) Doing race in the context of feminist interviewing: Constructing whiteness through talk. *Qualitative Inquiry*, vol. 9, no. 6, 895–914.

Blunt, A., Nash, F., Hatfield, M. and Sourch, C. (eds.) (2015) *Publishing and Getting Read. A Guide for Researchers in Geography*. London: Royal Geographical Society.

Bocking, S. (2011) Indigenous knowledge and the history of science. Race and colonial authority in Northern Canada. In A. Baldwin, L. Cameron, and A. Kobayashi (eds.),

Rethinking the Great White North: Race, Nature, and the Historical Geographies of Whiteness in Canada. Vancouver: UBC Press, 39–61.
Bravo, M. and Sörlin, S. (eds.) (2002) *Narrating the Arctic: A Cultural History of Nordic Scientific Practices*. Cambridge: Science History Publications.
Cameron, E. (2015) *Far Off Metal River: Inuit Lands, Settler Stories, and the Making of the Contemporary Arctic*. Vancouver: UBC Press.
Cardon, D. (2015) *À quoi rêvent les algorithmes. Nos vies à l'heure du big data*. Paris: Seuil/La République des idées.
Clark, T. (2008) "We're over-researched here!" Exploring accounts of research fatigue within qualitative research engagements. *Sociology*, vol. 42, no. 5, 953–970.
Collignon, B. (2004) Recueillir les toponymes inuit. Pour quoi faire? *Études/Inuit/Studies*, vol. 28, no. 2, 89–106.
Collignon, B. (2010) L'éthique et le terrain. *L'information géographique*, vol. 74, no. 1, 63–83.
Dahl, J. (1988) Self-government, Land Claim and Imagined Inuit Communities, *Folk*, vol. 30, 73–84.
Dassa, M., Kosmopoulos, C. and Pumain, D. (2010) JournalBase. Comparer les bases de données scientifiques internationales en sciences humaines et sociales (SHS). *Cybergeo: European Journal of Geography* [online], consulted October 2019. http://journals.openedition.org/cybergeo/22864
De Leeuw, S., Cameron, E., and Greenwood, M. (2012). Participatory and community-based research, Indigenous geographies, and the spaces of friendship: A critical engagement. *The Canadian Geographer/Le Géographe canadien*, vol. 56, n°2, 180–194.
Desrosières, A. (1993) *La politique des grands nombres. Histoire de la raison statistique*. Paris: La Découverte.
Desvaux, P. (2019) Introduction: les zabbâlîn, un objet sur-étudié? *Égypte/Monde arabe*, vol. 19, 9–32.
Eckert, D., Grossetti, M., Jégou, and Maisonobe M. (2014) Les villes de la science dans le monde. *Mappemonde*, vol. 4, no. 116.
Fabian, J. (1983) *Time and the Other: How Anthropology Makes Its Object*. New York: Columbia University Press.
Fenstermaker, S. and West, C. (eds.) (2002) *Doing Gender, Doing Difference: Inequality, Power, and Institutional Change*. Hove: Psychology Press.
Geertz, Clifford (2003) La description dense. Vers une théorie interprétative de la culture. In Daniel Cefaï (ed.), *L'enquête de terrain*. Paris: La Découverte, 208–333.
Giraut, F. and Houssay-Holzschuch, M. (2016) Place naming as dispositif: Toward a theoretical framework. *Geopolitics*, vol. 21, no. 1, 1–21.
Goodchild, M. F. and Hill, L. L. (2008) Introduction to digital gazetteer research. *International Journal of Geographical Information Science*, vol. 22, no. 10, 1039–1044.
Goodman, A., Morgan, R., Kuehlke, R., Kastor, S., Fleming, K., Boyd, J. and Aboriginal Harm Reduction Society (2018) "We've been researched to death": Exploring the research experiences of urban indigenous peoples in Vancouver, Canada. *The International Indigenous Policy Journal*, vol. 9, 2.
Graugaard, N. D. (2020) *Tracing Seal-Unsettling Narratives of Kalaallit Seal Relations*. Doctoral dissertation, Aalborg Universitetsforlag.
Grydehøj, A., Nadarajah, Y. and Markussen, U. (2020) Islands of indigeneity: Cultural distinction, indigenous territory and island spatiality. *Area*, vol. 52, no. 1, 14–22.
Harari-Kermadec, H. (2019) *Le classement de Shanghai. L'université marchandisée*. Lormont: Le Bord de l'eau.

Iliadis, A. and Russo, F. (2016) Critical data studies: An introduction, *Big Data & Society*, vol. 3, no. 2, 1–7.

Intergovernmental Panel on Climate Change (2018) *Global Warming of 1.5°C, Summary for Policy Makers*. Genève: IPCC.

Inuit Tapiriit Kanatami (2018) *National Inuit Strategy on Research*. Ottawa: Inuit Tapiriit Kanatami.

Jarvis, D. (2014) Regulating higher education: Quality assurance and neo-liberal managerialism in higher education—A critical introduction. *Policy and Society*, vol. 33, no. 3, 155–166.

King, R. (2009) *Governing Universities Globally: Organizations, Regulation and Rankings*. Cheltenham: Edward Elgar Publishing.

Kleivan, I. (1977) Place names in Greenland: Cultural imperialism and cultural identity. *Transactions of the Finish Anthropological Society*, vol. 2, 197–215.

Kleivan, I. (1996) Inuit oral tradition about Tunit in Greenland. In Bjarne Grønnow and John Pind (eds.), *The Paleo-Eskimo Cultures of Greenland: New Perspectives in Greenlandic Archeology*. Copenhagen: Danish Polar Center, 215–235.

Krupnik, I. (ed.) (2016) *Early Inuit Studies: Themes and Transitions, 1850s-1980s*. Washington, DC: Smithsonian Institution.

Larsen, J. N. and Fondahl, G. (eds.) (2015) *Arctic Human Development Report: Regional Processes and Global Linkages*. Copenhagen: Nordic Council of Ministers.

Lascoumes, P. and Le Galès, P. (2004) *Gouverner par les instruments*. Paris: Presses de Sciences Po.

Latour, B. and Wooglar, S. (2006) *La vie de laboratoire. La production des faits scientifiques*. Paris: La découverte.

Lefort, I. (2012) Le terrain: l'Arlésienne des géographes? *Annales de géographie*, vol. 687–688, no. 5, 468–486.

Louis, R. P. (2007) Can you hear us now? Voices from the margin: Using indigenous methodologies in geographic research. *Geographical Research*, vol. 45, no. 2, 130–139.

Maisonobe, M. (2015) *Étudier la géographie des activités et des collectifs scientifiques dans le monde: de la croissance du système de production contemporain aux dynamiques d'une spécialité, la réparation de l'ADN*. Ph. D. thesis, Toulouse Le Mirail University.

Maisonobe, M., Grossetti, M., Eckert, D., Jégou, L. and Milard, B. (2016) L'évolution mondiale des réseaux de collaborations scientifiques entre villes: des échelles multiples. *Revue Française de Sociologie*, vol. 57, no. 3, 415–438.

Matthiessen, C. W., Winkel Schwarz, A. and Find, S. (2002) The top-level global research system, 1997–99: Centres, networks and nodality. An analysis based on bibliometrics indicators. *Urban Studies*, vol. 39, nos. 5–6, 903–927.

Matthiessen, C. W., Winkel Schwarz, A. and Find, S. (2010) World cities of scientific knowledge: Systems, networks and potential dynamics. An analysis based on bibliometric indicators. *Urban Studies*, vol. 47, no. 9, 1879–1897.

Mermet, A-C. (2017) Airbnb and tourism gentrification: Critical insights from the exploratory analysis of the 'Airbnb syndrome' in Reykjavik. In M. Gravari-Barbas et S. Guinand (eds.), *Tourism and Gentrification in Contemporary Metropolises*. Abingdon: Routledge, 52–74.

Morin, F. (2001) La construction de nouveaux espaces politiques inuit à l'heure de la mondialisation. *Recherches amérindiennes au Québec*, vol. 31, no. 3, 25–36.

Neal, S., Mohan, G., Cochrane, A. and Bennett, K. (2016) "You can't move in Hackney without bumping into an anthropologist": Why certain places attract research attention, *Qualitative Research*, vol. 16, no. 5, 491–507.

Noucher, M. (2020) The place names of French Guiana in the face of the Geoweb: Between data sovereignty, indigenous knowledge, and cartographic deregulation. *Cartographica: The International Journal for Geographic Information and Geovisualization*, vol. 55, no. 1, 15–28.

Noucher, M., Hirt, I. and Arnauld de Sartre, X. (2019) Mises en chiffres, mise en cartes, mises en ordre du monde. *EspacesTemps.net* [En ligne], consulted June 2020. URL: www.espacestemps.net/articles/mises-en-chiffres-mises-en-cartes-mises-en-ordre-du-monde/

Nungak, Z. (2006) Introducing the science of Qallunology. *Windspeaker*, vol. 24, no. 2, 18–21.

Nuttall, M. (2010) Oil and gas development in the North: Resource frontier or extractive periphery? *The Yearbook of Polar Law Online*, vol. 2, no. 1, 225–243.

Pascucci, E. (2017) The humanitarian infrastructure and the question of over-research: reflections on fieldwork in the refugee crises in the Middle East and North Africa. *Area*, vol. 49, no.2, 249–255.

Petersen, R. (1984) The Pan-Eskimo movement. In D. Damas (ed.), *Handbook of North American Indians*, vol. 5: Arctic. Washington, DC: Smithsonian Institution, 724–728.

Petterson, C. (2014) *The Missionary, the Catechist and the Hunter. Foucault, Protestantism and Colonialism*. Chicago: Haymarket Books.

Petterson, C. (2016) Colonialism, racism and exceptionalism. In K. Loftsdóttir and L. Jensen (eds.), *Whiteness and Postcolonialism in the Nordic Region*. Abingdon: Routledge, 41–54.

Poppel, B. (ed.) (2015) *SLiCA: Survey of Living Conditions in the Arctic. Living Conditions and Quality of Life Among Inuit, Saami and Indigenous Peoples of Chukotka Peninsula*. Copenhagen: Nordic Council of Ministers.

Rose-Redwood, R. et Alderman, D. (2011) Critical interventions in political toponymy. *ACME: An International E-Journal for Critical Geographies*, vol. 10, no. 1, 1–6.

Rose-Redwood, R., Alderman, D. and Azaryahu, M. (2010) Geographies of toponymic inscription: New directions in critical place-name studies. *Progress in Human Geography*, vol. 34, no. 4, 453–470.

Sharp, C. et Murdoch, S. (2006) *How to Gather Views on Service Quality: Guidance for Social Landlords*. Edinburgh: Communities Scotland/Scottish Executive.

Simon, P. (2003) Les sciences sociales françaises face aux catégories ethniques et raciales. *Annales de démographie historique*, vol. 105, no. 1, 111–130.

Smith, L. T. (2008) *Decolonizing Methodologies: Research and Indigenous Peoples*. London: Zed Books.

Sontag, S. (1994) *The Anthropologist as Hero. Against Interpretation and other Essays*. London: Vintage.

Stoddart, M. and Smith, J. (2016) The endangered arctic, the arctic as resource frontier: Canadian news media narratives of climate change and the north. *Canadian Review of Sociology/Revue canadienne de sociologie*, vol. 53, no. 3, 316–336.

Sukarieh, M. and Tannock, S. (2013) On the problem of over-researched communities: The case of the Shatila Palestinian Refugee Camp in Lebanon. *Sociology*, vol. 47, no. 3, 494–508.

Thomsen, H. (1998) Ægte grønlændere og nye grønlændere—om forskellige opfattelser af grønlandskhed. *Den jyske historiker*, vol. 81, 21–69.

Titterton, M. and Smart, H. (2008) Can participatory research be a route to empowerment? A case study of a disadvantaged Scottish community. *Community Development Journal*, vol. 43, no. 1, 52–64.

Tuck, E. (2009) Suspending damage: A letter to communities. *Harvard Educational Review*, vol. 79, no. 3, 409–428.

Tuck, E. and Yang, K. W. (2012) Decolonization is not a metaphor. *Decolonization: Indigeneity, Education and Society*, vol. 1, 1–40.

Urry, J. (2003) La constitution du questionnaire notes and queries on Anthropology. Les premiers pas de l'anthropologie britannique 1870–1920. In Daniel Cefaï (ed.), *L'enquête de terrain*. Paris: La Découverte, 65–88.

Watson, A. E., Alessa, L. and Sproull, J. (2002) *Wilderness in the Circumpolar North: Searching for Compatibility in Ecological, Traditional, and Ecotourism Values*. Ogden: US Department of Agriculture, Forest Service.

Wilson, S. (2008) *Research is Ceremony: Indigenous Research Methods*. Halifax: Fernwood Publishing.

Zeller, S. and Ries, C. J. (2014) Wild men in and out of science: Finding a place in the disciplinary borderlands of Arctic Canada and Greenland. *Journal of Historical Geography*, vol. 44, 31–43.

9 Confessions of an 'academic tourist'

Reflections on accessibility, trust, and research ethics in the 'Grandhotel Cosmopolis'

Marielle Zill

Introduction

This chapter discusses the challenges and consequences of over-research in the context of asylum seeker accommodation by examining the case of the Grandhotel Cosmopolis (GHC) in Augsburg, Germany. The project is an asylum seeker centre as well as a tourist hotel and describes itself as a 'concrete utopia—realising a cosmopolitan everyday culture without limits where refugees, travellers, guests, artists and neighbours meet and are welcome' (Grandhotel Cosmopolis 2014). The project received multiple prizes and attracted scores of journalists, artists, and student researchers since its opening in 2013. The chapter critically examines fieldwork undertaken at the height of the 'refugee crisis' in Europe between September 2016 and July 2017, as part of a PhD project. During the so-called refugee crisis, 'doing something with refugees' became fashionable in the fields of journalism, research, and political art. 'Hot topics' such as 'refugee crisis' may result in 'over-research', meaning an excessive research focus on certain communities, projects, and places (Neal et al. 2016). The chapter argues that over-research is strongly related to temporary forms of research engagements pursuing 'hot topics' in places that are comparatively easy to access. The case study of the GHC highlights that over-research might not only produce research fatigue but the consequences of over-research on social relations between academia and local organisations and groups may be far greater and require both an individual and a collective effort to address the issue.

Over-research is particularly prominent in the field of refugee and migration studies, especially after the so-called refugee crisis in Europe in 2015. Despite the fact that research in the field of migration and refugee studies is often undertaken with the best of intentions, researchers do not always consider how the research process affects participants or whether these projects accord with their most pressing needs (Hugman et al. 2011; Jacobsen and Landau 2003). Following the high demand for knowledge on migration and refugees by both media and policymakers, the 'refugee crisis' gave birth to a 'refugee crisis industry' of which researchers are not only an important part but also are increasingly complicit with (Cabot 2019; Rozakou 2019; Stierl 2020). Being dependent upon humanitarian infrastructure to gain access to the

DOI: 10.4324/9781003099291-10

field, 'hot spots' have emerged for studying the plight of refugees, such as refugee camps in Jordan (Pascucci 2017), Lebanon (Sukarieh and Tannock 2012), or the infamous 'Moria' camp on Lesvos (Rozakou 2019). According to Cabot (2019), funding structures promote a form of 'crisis chasing', which reinforces mechanisms of over-research and conveys researchers a sense of status and authority by having studied a prominent hotspot. Likewise, Sukarieh and Tannock (2019) argue that a 'refugee research industry' is benefitting from the institutions and actors it is critiquing through its dependency on state funding and research agendas.

In the field of migration and refugee studies, over-research has several negative consequences for both research participants and the research process. In their study of the Shatila Palestinian refugee camp in Lebanon, Sukarieh and Tannock (2012) describe how previous research projects and documentary films had turned 'particularly promising' individuals into 'stars' which had negative long-term consequences for them and the larger community. Over-research in Shatila also led to co-dependencies between NGOs and camp residents, created an overly negative place image, and led to the commodification of research within the refugee camp. Relatedly, Pascucci (2017) found a kind of 'research savviness' on the side of participants in over-researched settings, which includes being well-informed about the research process and having higher expectations of research and its outcomes. Over-research also has negative consequences for the outcomes of research, one of which is called the 'streetlight effect'; a metaphor for how researchers tend to look for answers in places 'where the looking is good', rather than where the actual answers may be (Hendrix 2017). Similarly, over-research in the case of the Moria refugee camp contributed to its inaccessibility for researchers and journalists, while the high amount of knowledge produced on the topic mostly only served to reinforce its dystopian image 'of a place of destitution, abandonment and violence' (Rozakou 2019, p. 79).

After a short description of the Grandhotel Cosmopolis, the second section discusses the notion of academic tourism and its influence on the accessibility and positionality of the researcher. The third section describes how over-research affected relations of trust between the researcher and research participants in the GHC. Before concluding, the fourth section discusses strategies such as practising engaged reflexivity and knowledge co-production for addressing over-research.

The Grandhotel Cosmopolis

The Grandhotel Cosmopolis is a hotel, asylum seeker centre, café, restaurant, and artist and event space located in the inner city of Augsburg, Germany. From 2011 onwards, the former elderly care home was transformed by artists and activists into a project that calls itself a 'concrete utopia' (Grandhotel Cosmopolis 2016). The GHC is an art project inspired by the German artist Joseph Beuys and his concept of a 'social sculpture'; it is a 'societal artwork' in which 'everyone is welcome to participate' (Heber et al. 2011; Grandhotel Cosmopolis 2014, 2016). The first group of asylum seekers arrived in July 2013 and in October 2013 the project opened for hotel guests. The building has six floors, with a café/bar and hostel area on the ground floor, space for artists on the ground to third floor, rooms for 60 asylum seekers including shared kitchens and bathrooms on the first to third floor, 12 hotel rooms on the fourth and fifth floor, and a seminar room on the sixth floor. Public events are hosted in the café or in its restaurant located in the basement. The building is owned by the Protestant welfare organisation 'Diakonie', which rents the building to two parties: The non-profit association 'Grandhotel Cosmopolis e.V.' and the local district administration of Bavarian Swabia, who are responsible for housing asylum seekers. As described in an interview with the head of the welfare organisation, the local district administration had already prior to the idea of a 'grandhotel' expressed their interest in renting the building. By agreeing to the concept of an integrated hotel and asylum seeker centre, the number of asylum seekers to be accommodated in the building was reduced, which improved the overall quality of living for asylum seekers as it put less pressure on general facilities.

The project attracted significant local, regional, and national media attention, especially during the time of the so-called 'refugee crisis' in 2015 and won several regional and national prizes, such as the national 'Land of Ideas' competition (Grandhotel Cosmopolis 2016). A search in the news databank LexisNexis brings up over 100 results in German-speaking news media alone. Most major national newspapers, such as the weekly newspaper 'Die Zeit' and German national TV stations have reported on the project (Grandhotel Cosmopolis 2019). Its popularity also attracted a significant number of bachelor, master, and PhD students from all across Germany who wrote their thesis on the project, resulting in several publications (Costa Carneiro 2016; Marschall 2018). In contrast to other alternative accommodation centres such as Plan Einstein in Utrecht (Oliver et al. 2018), the GHC did not have a team of researchers responsible for a coordinated scientific assessment of the project and were relatively unprepared for the amount of media attention they received. The following section reflects on how over-research further complicated the process of gaining access to the GHC and how it challenged pre-conceived ideas on positionality in the field.

Playing the tourist: over-research as a consequence of 'academic tourism'?

The aim of the research project was to study how differences in the spatial, material, and institutional openness of asylum accommodation influenced contact and encounter between asylum seekers and neighbourhood residents (Zill et al. 2019). To this end, I had planned to conduct semi-structured interviews with both neighbourhood residents and asylum seekers living in the GHC, in combination with participatory observation. Having previous research experience in the GHC for my master's thesis, gaining access for me was easier as I benefitted from staying in contact with several members of the GHC. Researchers new to the project had taken a different approach; being denied initial access, one student researcher had booked a hotel room in the GHC and gained access 'as a tourist':

> Due to a lack of personal contacts and possible gatekeepers, contact was established via email, to which the response followed that there was not sufficient time to answer the request; in addition, it was stated that a participatory

> approach was central to the project. . . . Following the understanding that 'the ways into the field are as diverse as fieldwork itself', the researcher booked a room in the hotel and spent a week on site.
>
> (Fischer 2016, p. 53)

To be clear, the intent here is not to point fingers; rather, this excerpt reflects the common viewpoint that there are multiple ways of gaining access all of which have their advantages and disadvantages and are therefore equally valid approaches (Hammersley and Atkinson 2007). In other words, the strategies through which researchers obtain access may differ, yet their right to gain access is seldomly questioned in itself; as Bosworth and Kellezi (2016, p. 239) note, 'if it is discussed at all, is often cast as a one off arrangement, granted or withheld'. Taken individually, gaining access 'as a tourist' may not have immediate negative effects on the research setting. However, the collective impact of multiple researchers seeking to obtain permission without the explicit approval of an organisation may work to undermine trust between the researcher and the organisation. In the GHC, previous encounters with student researchers as well as the high number of requests by journalists had led to distrust between its members and researchers and journalists, which contributed to the higher social closedness of the setting. This social closedness, that is, the increasing difficulty of being granted access was supported by the viewpoint that a person is not a product that can be handed over for the purposes of data extraction. As stated repeatedly by several activists, the GHC 'is not a zoo for viewing refugees'. These statements echo more general critiques on the refugee research industry and the complicity of researchers with the processes and institutions they are critiquing (Cabot 2019; Sukarieh and Tannock 2019).

The extractive tendencies of academic research may be felt more strongly in places of over-research. Despite gaining access by securing the official approval to undertake research in the GHC, I was confronted early on with accusations of representing the 'university mentality' and with the implications of not conforming to unwritten codes of conduct. While I had agreed to do voluntary work, such as helping in the bar or hotel, along with providing translations, I was initially perceived as not active enough by one of the founders of the project, who accused me of 'playing the tourist':

> I started talking to Sarah about what she was doing, she asked what I was doing. We talked a bit about fieldwork and interviewing. Then Christian came and asked what I was doing, he claimed I was 'playing the tourist'. That I didn't know what I was doing and that he did not appreciate that. He said he does not like the University mentality, they just want to take things. The people in the Grandhotel were the ones doing something and what good is all that theory. . . . A part of me did feel attacked, and another part learned to not care and just take a note of it as a field observation. But I continued to feel tense, also not welcome and underappreciated to a certain extent.[1]
>
> (Fieldnotes, 25.10.2016)

In the GHC, over-research had contributed to an image of academic research as only serving its own interests and eroding societal trust in the university. 'Playing the tourist' is then a reference to a form of temporary and superficial engagement, similar to what Mackenzie et al. (2007) have described as 'fly in, fly out' research. The excerpt is a helpful starting point for reflecting on what is means to '*play the tourist*' in over-researched settings. First, 'playing the tourist' can be interpreted here as a kind of performance, as taking on a certain kind of role or habitus in this particular setting. Yet, this performance is not necessarily a conscious act; rather, the researcher is just as much produced by power relations within a setting. Following Gregson and Rose (2000, p. 441),

> performance—what individual subjects do, say, 'act out'—is subsumed within, and must always be connected to, performativity, to the citational practices which reproduce and subvert discourse, and which at the same time enable and discipline subjects and their performances.

Therefore, the first time I undertook research, I followed the GHC's rule to 'be active' and became a volunteer, resisting my researcher role. The second time, I felt pressured by time and project requirements to perform as 'the researcher'. In both cases, I reproduced one of the specific subject positions known to me. In the second case, my positionality akin to that of a tourist, which can be conceived as '*the academic tourist*'.

Academic research is itself a kind of performance, despite the widely held belief within academia that researchers are 'intentional, knowing, anterior subjects; able to interpret and represent a vast range of other social practices for academic audiences to interpret in turn, yet being themselves somehow immune from the same process; in other words, outwith academic power's script' (Gregson and Rose 2000, p. 447). The academic tourist then is a particular way of performing research activity, one that resembles the tourist performance several in the ways. Urry and Larsen (2011) outline several distinct characteristics generally associated with tourism, which can be employed to further define the notion of *academic tourism*. First, academic tourism involves 'movement of people to, and their stay in, various destinations' which are 'outside the normal places of residence and work. Periods of residence elsewhere are of short-term and temporary nature' (Urry and Larsen 2011, p. 4). In contrast to tourism, however, academic tourism may also target places that are close to a university or in other ways easily accessible. What still applies, however, is its characteristic to move somewhere and return, to be part of a setting for a short period of time, and constituting a form of *temporary* engagement. As postcolonial and Indigenous scholarship reminds us, going abroad, preferably to countries of the Global South, to undertake fieldwork is and always has been a privilege accorded to universities of the Global North (Bhambra 2013; Smith 2013). These uneven privileges have their history in the formation of disciplines themselves and their involvement in colonialist enterprises; following Tilley (2017, p. 27), 'the systemic extraction of raw commodities from (formerly) colonised countries finds its analogue in academics' piratic practices of "raw" data extraction for processing into refined intellectual property, to be published at prices which exclude the original contributing "knowers" '.

Second, academic tourism likewise involves a selection of certain kind of places which are hyped or are associated with certain desires, pleasures, or fantasies and are in some way 'out of the ordinary' (Urry and Larsen 2011, p. 4). The images of these places may similarly be projected not only via the media but also through academic publications and policy briefings. Of particular interest to academics studying marginalised communities are thus places with images of danger or precarity, from the classic 'ghetto' to modern day favelas, border zones, and refugee camps (Pascucci 2017; Rozakou 2019). Interestingly, significant overlaps are emerging between academic and conventional forms of tourism through the development of volunteer tourism or war-zone tourism (Mostafanezhad 2013; Mahrouse 2016). Third, the academic tourist gaze is also built upon practices of signification; whereas tourists might look for signs of what they regard as typical local behaviour, the academic tourist is also searching for people, cases, or materials that are informed by a particular idea or theory. Schlosser (2014, p. 203) is critical of an academic gaze informed by empiricist epistemology, which takes for granted a hierarchical relationship between theory, method, and the field. Instead, he argues for a reflexive research practice in which the field informs theory and researchers of what is 'unknown, unknowable, or situationally contingent.' Finally, similar to the effects of mass tourism, academic tourism may contribute to 'new socialised forms of provision . . . to cope with the mass character of the gaze of tourists' (Urry and Larsen 2011, p. 4), as has already been shown for refugee camps in Lebanon (Sukarieh and Tannock 2012, 2019). The next section will discuss relations of trust with asylum seekers, along with the possible effects of temporary research engagements.

Trust is like a crocodile: over-research and project-based contact with asylum seekers

Over-research contributed to changes in the norms and rules of conduct in the GHC, such as being critical of the practice of referral of an asylum seeker by a gatekeeper for the purpose of an interview. Therefore, the more 'conventional' approach of using gatekeepers proved nearly impossible. To gain social access, it was necessary to participate and be active within the GHC. Yet, being a young, white, and German woman volunteering in the project made me, for most interviewees, part of the group of 'activists'. As Karim,[2] an asylum-seeking resident and long-term volunteer in the GHC explained, some resident asylum seekers were afraid that supporting the goals of the activists might have a negative impact on their asylum procedure. Consequently, being perceived as an activist meant that I potentially received a similar level of mistrust as demonstrated by the following excerpt. Karim refuses to act as a gatekeeper for gaining access to other prospective respondents, as trust is not established by momentary smile or friendly facade but has to be established over longer periods of time. Over-research may exacerbate feelings of distrust towards researchers, in particular, with 'hard-to-reach' groups such as asylum seekers and refugees which have become a 'hot topic' for research and policy interventions (Stierl 2020).

> 'There is a lack of trust between the refugees and the activists and they need to do things to re-establish their trust. You need to build things together, work together to establish trust, do activities, cook together. But one problem is that the refugees are dependent on the system, as they want to enter normal society and leave the centre. So they distrust the activists, because they are against the system, some of them think that too much involvement with the activists might hurt their future chances. Some of them might even think that they are connected to the police, as they come from places where the system worked like that, so they are distrustful'. . . . Later I asked him if he knew anyone I could talk to. He said, what good would it do for him to introduce me to someone? Trust is like a crocodile—[pulling his face into a broad smile] I need to talk to people to gain their trust.
>
> (Fieldnotes, 26.05.2017)

Not only does over-research increase the levels of distrust by heightening the experiences of lack of impact and temporariness of research but also it may affect what is being said and what is left silent, thus influencing the quality of data gathered. Over-research contributes to already existing difficulties in establishing trust and rapport when studying refugees or irregular migrants (Hynes 2003; Níraghallaigh 2014). As Hynes (2003) highlights, all stages of being a refugee are characterised by mistrust. This high degree of mistrust within different stages of the refugee experience does not automatically mean that research on refugees is impossible; rather, researchers need to be aware of the potential for mistrust. Therefore, 'we need to choose whether we research *for, on* or *with* refugees' (Hynes 2003, p. 14). During my interviews with resident asylum seekers, I felt a level of discomfort I could not explain. I confided in Ahmed,[2] a resident asylum seeker whom I trusted and spent a lot of time with. His reply, presented in the following fieldnote excerpt, indicates that over-research also affects the kind of data researchers gather. Research, in the eyes of participants, becomes less a way of translating experiences than an end in itself within the 'refugee research industry' (Sukarieh and Tannock 2019). As a consequence, researchers' conversations with respondents may 'separate from heart and truth':

> I told him about my struggles in talking to people, he said, 'you can talk a lot to people but they will not talk to you with their heart, don't you feel that? They have conversations that are separate from heart and truth'. 'Yes, I do' I said, that is where the discomfort comes from. 'Also they see me as someone from the team'. 'How could they not? We cannot escape our positions'.
>
> (Fieldnotes, 17.05.2017)

Researchers finding themselves in situations of over-research need to take unequal power-relations between themselves and their research participants into account and how they intentionally or unintentionally exploit these power-relations. In order to 'collect data', researchers are trained to develop rapport with research participants, to show empathy when they feel none, or to 'fake friendship' with people they would under other circumstances not have considered 'friends' (Oakley 1981). Over the course of research, participants may develop expectations of friendship, especially those that do not have a large social support system. Despite the fact that formal consent is obtained, in practice, participants' contextual realities may limit them in their capacity to provide consent (Thompson 2002). This is already problematic under 'normal' circumstances and yet becomes a profound ethical dilemma in situations of over-research, especially in situations where vulnerable groups such as asylum seekers and refugees are involved (Mackenzie et al. 2007). Amooz, a young, male asylum seeker from Afghanistan and volunteer in the GHC explained that he disliked the pretence involved when refugees are approached on the basis of a project. He argued that while he appreciated help, it should be based in real interest in friendship and an understanding of mutuality. This is captured in his wish to be invited into someone's home to establish 'real' relationships:

> For example, when somebody wants to help refugees voluntarily, that's okay. Helping, accompanying, but not because of a shitty project, because they want to finish a project. And then they say, bye. They don't want to know you. . . . Project is finished, they leave. . . . It's also okay if they do a project. But not come to you because of the project, to say hello. It would be cool if also when there is no project, that he says hello. For example, taking me to his home and live together without a project, hey, how are you.[3]
>
> (Interview with Amooz)

In summary, over-research of vulnerable groups such as asylum seekers may lead to considerable ethical difficulties regarding the establishment of trust and rapport. It is crucial not to frame research participants as victims of researchers, as they choose to participate due to certain expectations emerging from this encounter, such as help with translations and emotional or other kinds of support (Mackenzie et al. 2007). More importantly, however, over-research in the form of high numbers of researchers with a temporary stay may worsen feelings of loss and cause considerable emotional harm to individuals with limited or fragile social networks. The next section will return to the notion of academic tourism and reflect on different strategies to process and approach over-research.

From academic tourist to academic-in-residence: strategies to address the consequences of over-research

Over-research is first and foremost a question of research integrity, yet one that still has to be recognised as such. According to Kaiser (2014, p. 341), research integrity is defined as a situation in which 'its practitioners behave in accordance with the accepted rules of good conduct within that system'. The problem regarding over-research is that as of now there are no ethical and methodological standards in place for defining 'good conduct' in situations of over-research. We need to differentiate here between our individual and our collective responsibilities towards research integrity. This section addresses the question of individual responsibility based on insights gained from a post-fieldwork engagement as an 'academic-in-residence' in the GHC.

Adopting a practice of engaged reflexivity: acknowledging the academic tourist in me

To gain an understanding of the dynamics and implications of over-researched situations, what we as individual researchers should reflect upon is our relations with others, as our subject positions are constituted by these everyday interactions. Research, in this understanding, is 'a process of constitutive negotiation' (Rose 1997, p. 316). In my struggle to uphold my performance as 'the researcher', I felt a sense of discomfort I could not explain. It is the awareness of and will to engage with this discomfort which prompted me to recognise the specific inter-personal dynamics characteristic to over-researched places. A first step in addressing our individual responsibilities towards over-research is then to practise reflexivity and 'engaged self-critique' (Cabot 2019). As feminist geographers have argued, reflexivity has its challenges and limitations. Particularly problematic is the notion of 'transparent reflexivity', which assumes that as researchers we are capable of fully grasping the landscapes of power in which we are operating and our positionality within them (Rose 1997). Despite these challenges, however, there are different kinds of reflexive practices which nevertheless constitute helpful tools for detecting and understanding situations of over-research.

Researchers in over-researched settings are often faced with research fatigue, which is expressed as apathy or indifference towards engagement in research projects (Clark 2008). Researchers' ability to determine when a situation is 'over'-researched therefore necessitates that individual researchers actively engage with the emotional landscapes of the places and cases they are studying. Being reflexive of our own emotions thus constitutes one of the tools to detect and understand the inter-personal dynamics of over-research (Davidson et al. 2007). Frequently, emotions are associated with a failure in 'neutrality' and 'objectivity', with possible consequences for one's future career (Widdowfield 2000). However, researchers not engaging with emotional experiences during fieldwork in the worst case run the risk of doing emotional harm to both themselves and their research participants and at best are neglecting a potentially enlightening field of knowledge. This is built on the understanding that emotions are relational; as Widdowfield (2000, p. 200) states, 'not only does the researcher affect the research process but they are themselves affected by this process'. Emotional reflexivity is therefore key to detecting and understanding over-researched settings, as research fatigue and distrust are not always openly voiced but may surface in the behaviour of those we engage with during the research process. In the words of respondent Ahmed, research is often undertaken 'separate from heart and truth'. Practising emotional reflexivity and understanding how our own emotions are tied up with those of others may thus help to detect and understand the emotional landscapes of over-researched settings and how these may influence the data collected.

Practically speaking, writing down and reflecting upon the feelings which we think we should not feel and certainly do not publicly want to acknowledge is a first step towards detecting and understanding the situations of over-research. The aforementioned fieldnote excerpts and interview quotes exhibit the relational nature of feelings trespassing between the activists, asylum seekers and myself, such as feeling underappreciated and unwelcome when confronted with the accusation of 'playing the tourist'. One way of learning from the emotional landscapes of over-researched settings is then to not only pay attention to how one feels but also to our own moral judgements about those feelings. Emotions tend to be noticed when they run up against so-called 'feeling rules' (Young and Lee 1996) of how and what one 'ought' to feel during fieldwork. As Bondi (2007, p. 236) notes, 'the co-construction of data in interpersonal relationships requires both researchers and those with whom they interact to deploy a wide range of skills to which emotional life is integral'. Consequently, Bondi (2007) argues that researchers should have support structures to analyse feelings within their research community, as feelings can be easily misinterpreted (Bondi 2007). Moreover, given the relational nature of emotions, neglecting our emotional life may affect our ability to do research, as well as influence the way we relate to research participants. The following section discusses the possibilities and limits of 'relating differently' with over-researched settings.

Relating differently: from collecting data to collective data?

In the following, two other accusations regarding 'academic tourism' are addressed: First, the extractive manner of research 'taking things' and second, the usefulness of theoretical abstraction or 'what good is all that theory'. Along with discomfort, I felt a sense of failure that arose out of the conviction that a different way of relating with the field was necessary to uphold research integrity. As described in section three, as an academic tourist I was seen to embody a 'university mentality' of 'just taking things'. This is a critique of the extractive character of research, which is often felt more strongly in situations of over-research and echoes criticism levied against mass or 'over-tourism' (Seraphin et al. 2018). Feminist and postcolonial scholars in particular have criticised the extractive nature of research, especially when knowledge is expropriated from the Global South and fed into the knowledge circuits of the global north (Jazeel and McFarlane 2010; Halvorsen 2018). In the case of 'over-tourism', general recommendations are to find a form of management that either restricts or bans tourism altogether or to develop a form of sustainable tourism for each particular location (Borg et al. 1996; Russo 2002). To address the extractive nature of research and develop more sustainable research practices, it is necessary to not only *think about* but also *practice* a form of research that takes its *collective* impact into account.

Honestly addressing over-research and academic tourism requires not only different methodologies but also a critical interrogation of our research ethics. We should be highly critical of 'easy fixes' to over-research that address methodology alone, such as calls for more participatory approaches. As Pain and Francis (2003, p. 53) remind us, 'the term "participatory" should be avoided when the primary intention is traditional "extractive" research for the purposes of gathering information'. A change of methods is therefore not sufficient to address over-research as this problem concerns not only the way we select research topics and field sites but also which kinds of relations we want to engage in and sustain with the people and communities we study. Responsible academic research for over-researched places then requires honesty about our own intentions of doing research towards both ourselves, our research participants, and the academic community. To not *only to take* but also *to give* we need to engage with the possibilities and limitations of reciprocity. It means not to shy away from asking why and for whom we are doing this research and who will *truly* benefit from it. As outlined earlier, a relational form of reflexivity is a part of this, along with a serious engagement with the politics of knowledge production (Jazeel and McFarlane 2010; Routledge and Derickson 2015). Wherever possible, this means resisting institutional pressures towards academic tourism, characterised by short-term forms of research engagements and the temptations of hot topics, as these may result in over-research, as exemplified by research on refugee camps (Sukarieh and Tannock 2012, 2019).

Giving back, that is, engaging in reciprocal relations is influenced by our ontological and epistemological assumptions towards our research subjects. Ontologically, this means to question who we see as producers of knowledge and whether knowledge is created *on* or *with* our research subjects. Scholar-activists have claimed that social movements are often the basis for theoretical innovations and shape academic knowledge production in profound ways. Social movements should therefore be seen as 'knowledge producers in their own right', rather than mere 'objects of knowledge' (Chesters 2012, p. 153). Any efforts of researchers to position themselves at a distance or as 'an observer' may have a negative impact not only on relations of trust, as outlined earlier, but also on reciprocity. The GHC produces practical, inter-subjective knowledge by establishing an alternative form of asylum accommodation. Its members have been invited to speak in forums and conferences all across Germany on this topic, while the academic debate on alternative forms of accommodation has followed much later. Creating knowledge together *with* projects such as the GHC requires an awareness of the specific questions of and a close connection with the local level. While this might not always be possible, joint knowledge production together with the subjects of our research might lead our research to be more current—not being attracted to a topic when it is already 'hot' and to be there before it becomes 'hot'. Engaging in joint knowledge production could also help avoid research fatigue, even in places which receive a lot of research attention, as research fatigue is caused by not being in tune with the questions and issues 'on the ground' and by insufficiently striving towards reciprocal relationships with the subjects of our research.

Addressing over-research requires moving from 'collecting data to collective data', meaning a more responsible form of knowledge production which includes accountability towards the local context in which our research is situated. To learn how to 'relate differently' and to 'give back', I returned to the GHC towards the end of my research project to become an 'academic-in-residence'. Between the months October 2019 and January 2020, I rented a desk in the GHC with some leftover research funding, dividing my time between finishing up my academic writing and helping out with whatever was needed in the everyday running of the project. In November 2019, I organised a public event in which I and two other speakers presented our academic findings related to innovative forms of asylum accommodation. These four months gave me a glimpse of what it means to 'relate differently' and how academic knowledge can be made useful in an activist context. Knowledge co-production is not necessarily about a particular method, but about making knowledge production more transparent, accessible and open to forms of responsible learning (Jazeel and McFarlane 2007). As an 'academic-in-residence', I found myself inserting theoretical insights, concepts and findings of my own work into everyday conversations. Not in the form of a lecture, but in dialogue as a way to give a name to on-going structures and processes. Some of these 'theory snippets' echoed back when they proved useful for clarifying problems at hand, teaching me which theoretical lenses might constitute tools for social change. More than any particular method, it was my daily presence and my long-term engagement with the GHC that created the conditions and relationships for dialogue. Moving towards joint forms of knowledge production does not mean to do away with abstraction; instead, we need to inquire 'how knowledge produced through research might be of use to multiple others without re-inscribing the interests of the privileged; and how such knowledge might be actively tied to a material politics of social change' (Routledge and Derickson 2015, p. 393). While there are different strategies for how this can be achieved, theory and knowledge production can be made useful when it is accountable to its context and produced in dialogue. In an increasingly complex world, it is not only our privilege but our task to make the process and products of abstraction publicly available.

Addressing academic tourism: a question of research ethics in over-researched places

This chapter discussed the challenges of over-research in the context of migration and refugee research by examining fieldwork undertaken in an innovative form of asylum seeker accommodation, the Grandhotel Cosmopolis in Augsburg, Germany. During the 'refugee crisis', this project attracted scores of journalists, students, and researchers, which led to over-research and research fatigue among its inhabitants and members of staff. Over-research also led to challenging interpersonal dynamics, such as difficulties in gaining access to research participants due to the lack of trust and the increased social closedness of the setting. Social closedness resulted from the contention that 'refugees are not a product' for research, thus, closedness emerged to prevent the commodification of research and refugees. In addition, commodified social relations in the form of 'project-based' contact may take advantage of individuals in marginalised positions with limited social support systems and lead to feelings of loss. The case is illustrative of larger dynamics within academic knowledge production, such as 'crisis chasing', motivated by funding structures and public pressures to research 'hot topics' (Cabot 2019).

Over-researched places should not be seen as exceptions to the norm but rather as a magnifying glass for the norm. As researchers, we bear responsibility not only for our individual, but also for our collective performances and their consequences. Over-research and research fatigue are not marginal phenomena but may constitute one of the greatest challenges social science scholars have to face in the upcoming decade. Given the growth in student numbers in higher education and in research projects across the globe, an intensification of over-research is to be expected. It is crucial then that we do not shy away from interrogating uncomfortable or disorienting moments, such as being accused of 'academic tourism'. Reflecting, rather than shying away from our own emotions, may constitute a first step in acknowledging that 'something is not quite as it should be'. Beyond the individual research encounter, over-research may influence the relationship between university and society; changes in this relationship are already mirrored in increasing pressures of societal impact assessments (Pain et al. 2011). Similarly, under-research may also be undesirable as the places, cases, or communities are neither represented within our findings nor can they be considered in policymaking (Omata 2019).

What can researchers do to address over-research? In short, the credo is 'beware and be aware'; beware of 'hot topic' research, famous or hyped places, cases or communities, of your own 'good intentions', and your desire to set yourself apart. Beware also of 'easy fixes' to over-research, such as calls for more participatory methods, as these do not necessarily change the extractive character of research itself. Over-research is first and foremost not a question of methodology but of research ethics. At its core lies the question of how we choose to relate towards our research subjects and objects. Over-research therefore demands both an individual and a collective response; it requires individual awareness and a collective effort to engage with and address inequalities in the current system of academic knowledge production. Still, every collective shift starts with individual awareness: Be aware that researchers have come before you and will come after you. Be aware of the places, cases, or communities that are flying under the radar for they also have stories to tell. Be aware that social movements, collectives, and communities are also producers of knowledge and that they too are 'experts'. Be aware that while the products of academic knowledge may not be of interest to all, this does not mean that the process of abstraction and search for explanations of complex realities may still interest 'non-academics'. Be aware that your status as an academic comes with both privileges and duties; especially academics of the Global North are afforded privileged access to resources and education. It is our duty to reflect on this privilege and use it not only towards contributions to theory and knowledge but also to engage in the co-production of knowledge whenever and wherever feasible. Finally, be aware of your own emotions and use them to critically interrogate the individual and collective dynamics of knowledge production. Being honest with ourselves and our research participants about the limitations of our research might seem daunting but just as well might establish a solid foundation for re-energizing the relationship between academia and society.

Notes

1 All fieldnotes were originally written in English.
2 Pseudonym. All respondents were anonymised for the purpose of research.
3 Translated from German.

References

Bhambra, G.K. (2013) The possibilities of, and for, global sociology: A postcolonial perspective. *Political Power and Social Theory*, 24, 295–314.

Bondi, L. (2007) The place of emotions in research: From partitioning emotion and reason to the emotional dynamics of research relationships. *In*: J. Davidson, L. Bondi, and M. Smith, eds. *Emotional Geographies*. Aldershot: Ashgate, 231–246.

Borg, J. Van Der, Costa, P., and Gotti, G. (1996) Tourism in European heritage cities. *Annals of Tourism Research*, 23 (2), 306–321.

Bosworth, M., and Kellezi, B. (2016) Getting in, getting out and getting back: Conducting long-term research in immigration detention centres. *In*: S. Armstrong, J. Blaustein, and A. Henry, eds. *Reflexivity and Criminal Justice: Intersections of Policy, Practice and Research,* London: Palgrave Macmillan UK, 1–388.

Cabot, H. (2019) The business of anthropology and the European refugee regime. *American Ethnologist*, 46 (3), 261–275.

Chesters, G. (2012) Social movements and the ethics of knowledge production. *Social Movement Studies*, 11 (2), 145–160.

Clark, T. (2008) 'We're over-researched here!': Exploring accounts of research fatigue within qualitative research engagements. *Sociology*, 42 (5), 953–970.

Costa Carneiro, J. (2016) Grandhotel Cosmopolis als Ort der Bildung von Gesellschaft. Spannungen zwischen Utopie und Wirklichkeit. *In*: M. Ziese and C. Gritschke, eds. *Geflüchtete und Kulturelle Bildung. Formate und Konzepte für ein neues Praxisfeld.* Bielefeld: Transcript, 313–324.

Davidson, J., Bondi, L., and Smith, M. (2007) *Emotional Geographies*. Aldershot [etc.]: Ashgate.

Fischer, R. (2016) *Unterbringung Geflüchteter neu denken.* Genese einer sozialen Plastik als Wohn- und Sozialraum: Das Grandhotel Cosmopolis. Universität Osnabrück.

Grandhotel Cosmopolis (2014) *Concept Grandhotel Cosmopolis* [online]. Available from: https://grandhotel-cosmopolis.org/wp-content/uploads/2014/06/Grandhotel-Erst-Konzept_2011.pdf [Accessed 29 October 2019].

Grandhotel Cosmopolis (2016) Infos zum Grandhotel Cosmopolis—Stand 5. Juli 2016, 1–6.

Grandhotel Cosmopolis (2019) *Pressearchiv* [online]. Available from: https://grandhotel-cosmopolis.org/de/category/presse/ [Accessed 11 June 2019].

Gregson, N., and Rose, G. (2000) Taking Butler elsewhere: Performativities, spatialities and subjectivities. *Environment and Planning D: Society and Space*, 18 (4), 433–452.

Halvorsen, S. (2018) Cartographies of epistemic expropriation: Critical reflections on learning from the south. *Geoforum*, 95 (June), 11–20.

Hammersley, M., and Atkinson, P. (2007) Access. *In*: *Ethnography: Principles in Practice*, 3rd ed., Abingdon: Routledge.

Heber, G., Adamczyk, M., and Kochs, S. (2011) *Konzept für eine soziale Skulptur in Augsburgs Herzen* [online]. Available from: https://grandhotel-cosmopolis.org/wp-content/uploads/2014/06/Grandhotel-ErstKonzept_2011.pdf [Accessed 17 May 2019].

Hendrix, C.S. (2017) The streetlight effect in climate change research on Africa. *Global Environmental Change*, 43, 137–147.

Hugman, R., Pittaway, E., and Bartolomei, L. (2011) When 'do no harm' is not enough: The ethics of research with refugees and other vulnerable groups. *British Journal of Social Work*, 41, 1271–1287.

Hynes, P. (2003) *The Issue of 'Trust' or 'Mistrust' in Research with Refugees: Choices, Caveats and Considerations for Researchers*. New Issues in Refugee Research. Geneva.

Jacobsen, K., and Landau, L.B. (2003) Researching refugees: Some methodological and ethical considerations in social science and forced migration. *New Issues in Refugee Research*, 27 (90), 27.

Jazeel, T., and McFarlane, C. (2007) Responsible learning: Cultures of knowledge production and the north-south divide. *Antipode*, 39 (5), 781–789.

Jazeel, T., and McFarlane, C. (2010) The limits of responsibility: A postcolonial politics of academic knowledge production. *Transactions of the Institute of British Geographers*, 35 (1), 109–124.

Kaiser, M. (2014) The integrity of science—Lost in translation? *Best Practice and Research: Clinical Gastroenterology*, 28 (2), 339–347.

Mackenzie, C., Mcdowell, C., and Pittaway, E. (2007) Beyond 'Do no harm': The challenge of constructing ethical relationships in refugee research. *Journal of Refugee Studies*, 20 (2), 299–319.

Mahrouse, G. (2016) War-zone tourism: Thinking beyond voyeurism and danger. *Acme*, 15 (2), 330–345.

Marschall, A. (2018) What can theatre do about the refugee crisis? Enacting commitment and navigating complicity in performative interventions. *Research in Drama Education*, 23 (2), 148–166.

Mostafanezhad, M. (2013) The politics of aesthetics in volunteer tourism. *Annals of Tourism Research*, 43, 150–169.

Neal, S., Mohan, G., Cochrane, A., and Bennett, K. (2016) 'You can't move in Hackney without bumping into an anthropologist': Why certain places attract research attention. *Qualitative Research*, 16 (5), 491–507.

Níraghallaigh, M. (2014) The causes of mistrust amongst asylum seekers and refugees: Insights from research with unaccompanied asylum-seeking minors living in the republic of Ireland. *Journal of Refugee Studies*, 27 (1), 82–100.

Oakley, A. (1981) Interviewing women: A contradiction in terms. *In*: H. Roberts, ed. *Doing Feminist Research*. London: Routledge and Kegan Paul, 30–61.

Oliver, C., Geuijen, K., and Dekker, R. (2018) Local innovation to overcome the challenges of asylum seeker reception. The Utrecht Refugee Launchpad.

Omata, N. (2019) 'Over-researched' and 'under-researched' refugees. *Forced Migration Review*, (June), 15–18.

Pain, R., and Francis, P. (2003) Reflections on participatory research. *Area*, 35 (1), 46–54.

Pain, R., Kesby, M., and Askins, K. (2011) Geographies of impact: Power, participation and potential. *Area*, 43 (2), 183–188.

Pascucci, E. (2017) The humanitarian infrastructure and the question of over-research: Reflections on fieldwork in the refugee crises in the Middle East and North Africa. *Area*, 49 (2), 249–255.

Rose, G. (1997) Situating knowledges: Positionality, reflexivities and other tactics. *Progress in Human Geography*, 21 (3), 305–321.

Routledge, P., and Derickson, K.D. (2015) Situated solidarities and the practice of scholar-activism. *Environment and Planning D: Society and Space*, 33 (3), 391–407.

Rozakou, K. (2019) 'How did you get in?' Research access and sovereign power during the 'migration crisis' in Greece. *Social Anthropology*, 27 (S1), 68–83.

Russo, A.P. (2002) The 'vicious circle' of tourism development in heritage cities. *Annals of Tourism Research*, 29 (1), 165–182.
Schlosser, K. (2014) Problems of abstraction and extraction in cultural geography research: Implications for fieldwork in Arctic North America. *Journal of Cultural Geography*, 31 (2), 194–205.
Seraphin, H., Sheeran, P., and Pilato, M. (2018) Over-tourism and the fall of Venice as a destination. *Journal of Destination Marketing and Management*, 9 (September 2017), 374–376.
Smith, L.T. (2013) *Decolonizing Methodologies: Research and Indigenous Peoples*. 2nd ed. London: Zed Books Ltd.
Stierl, M. (2020) Do no harm? The impact of policy on migration scholarship. *Environment and Planning C: Politics and Space*, 0 (0), 1–20.
Sukarieh, M., and Tannock, S. (2012) On the problem of over-researched communities: The Case of the Shatila Palestinian refugee camp in Lebanon. *Sociology*, 47 (3), 494–508.
Sukarieh, M., and Tannock, S. (2019) Subcontracting academia: Alienation, exploitation and disillusionment in the UK overseas Syrian refugee research industry. *Antipode*, 51 (2), 664–680.
Thompson, S. (2002) My research friend? My friend the researcher? My friend, my researcher? Mis/informed consent and people with developmental disabilities. *In*: W.C. Hoonaard, ed. *Walking the tightrope. Ethical issues for qualitative researchers*. Toronto: University of Toronto Press, 95–106.
Tilley, L. (2017) Resisting piratic method by doing research otherwise. *Sociology*, 51 (1), 27–42.
Urry, J. and Larsen, J. (2011) *The Tourist Gaze 3.0*. Theory, culture & society. London: Sage.
Widdowfield, R. (2000) The place of emotions in academic research. *Area*, 32 (2), 199–208.
Young, E. and Lee, R. (1996) Fieldworker feelings as data: 'emotion work' and 'feeling rules' in the first person accounts of sociological fieldwork. *In*: V. James and J. Gabe, eds. *Health and the Sociology of Emotions*. Oxford: Blackwell, 97–114.
Zill, M., van Liempt, I., Spierings, B., and Hooimeijer, P. (2019) Uneven geographies of asylum accommodation: Conceptualizing the impact of spatial, material, and institutional differences on (un)familiarity between asylum seekers and local residents. *Migration Studies*, 0 (0), 1–19.

10 Locating climate change research

The privileges and pitfalls of choosing over- and under-researched places

Chandni Singh

Introduction

Where one does research is a critical yet less discussed aspect of research methodology. In climate change vulnerability and adaptation studies especially, where risks and the strategies people use to manage them are extremely contextual, the 'research site' has tremendous bearing on what one finds and the recommendations one draws out. In this chapter, I use examples of climate change adaptation studies across rural and urban India to highlight how there are pros and cons of both over- and under-researched places. There is no one 'right' place to conduct research, however, recognising the biases and benefits that a location signifies is critical for more reflexive research to not only build upon existing work but also give voice to silences in places that are under-researched. This, I argue, is especially important for climate change research in rapidly changing and highly vulnerable countries such as India.

All research sites are not the same. Some have a history of being researched, replete with longitudinal datasets and theorisation from them, while others are conspicuous by their absence (as highlighted by Hanna A. Ruszczyk's chapter in this volume; Chapter 7). In any research, choosing a research site is deeply personal, often challenging, and carries with it expectations of having to defend choices around why a certain site was chosen. Common challenges associated with choice of research locations are issues of generalizability (e.g. are the findings too geographically niche?); insider/outsider dilemmas (e.g. balancing biases that might creep into being too close to one's respondents or oversights in data collection by being too unfamiliar with social norms); and representativeness (e.g. does the site represent existing variability?).

In climate change adaptation research especially, the unprecedented and urgent nature of the problem deepens the burden placed on choice of research location. The choice of location in climate change research is a proclamation of who is acknowledged as vulnerable and which places are identified as exposed to risk. Moreover, in climate change research, temporality (change over time) is a key aspect of understanding climate risks and responses, and how we can adapt in the future (Fawcett et al., 2017), which has led to calls for conducting longitudinal

DOI: 10.4324/9781003099291-11

and/or repeated research in sites to understand long-term change and pathways of climate resilient development.

Within India, the focus of this chapter, climate change researchers have tended to fall back on certain usual suspects, that is, places that have been previously studied by development researchers, or places repeatedly identified as at very high risk. For example, climate migration studies tend to focus on the Sunderbans in eastern India, which is highly exposed to visible, rapid-onset extreme events such as flooding and inundation. This has meant that other geographies witnessing climate-driven migration such as the semi-arid regions of Maharashtra and Karnataka, or Himalayan towns of northern India have received relatively less research attention (in the climate adaptation literature).

Conservatism in choosing research locations, that is, the practice on falling back on places with high risk or strong adaptation action, can reinforce data from particular sites. This repeated focus on certain locations, while important, potentially takes away attention from understudied yet similarly high-risk places and can reinforce geographical biases. For example, exposure to and impacts of sea level rise in India regularly cite examples from megacities such as Mumbai, Kolkata, and Chennai, while research on highly exposed medium-sized cities on the eastern coast such as Visakhapatnam or Bhubaneshwar remains relatively lower. This focus on metropolitan cities is understandable—these cities are home to populations greater than 10 million people and concentrate risk through their built infrastructure, existing socio-economic vulnerability, and highly exposed coastal geography. However, this megacity bias is conservative in a country like India where medium- and small-sized cities (relatively large by population size) are experiencing higher risks, often with less financing and attention to develop climate action plans. Notably, 'small cities accumulate disaster risk in similar ways to larger urban centres, but without the attendant growth in infrastructure or governance capacity' (Rumbach, 2016, p. 109), making the case for *more*, not less research focus in them. This conservatism also alludes to a deeper challenge of knowledge creation in climate research—one of relying on iconic examples to provide solutions that might not be suited to smaller cities and settlements (Bulkeley, 2006).

The tendency of being locationally conservative means that fast-changing spaces—peri-urban areas, small towns, understudied villages—are often ignored (Singh et al., 2017). Given that many of these under-researched places are also climate hotspots, that is, 'regions that are particularly vulnerable to current or future climate impacts, and where human security may be at risk' (de Sherbinin, 2014, p. 23) this is a critical omission. Also, places with knowledge gaps continue to fall off maps, with the lack of baseline data and existing knowledge potentially deterring new research. This can reinforce the bias towards over-researched places.

In this chapter, I first review the importance of place in climate change research and then draw on examples from rural and urban India to examine the pitfalls and privileges of choosing over- and under-researched places for climate change vulnerability and adaptation research. I end with a discussion on what implications choice of location might hold for research methodology and climate action.

'Place' in climate change research

The role of place in climate impacts, risks, and adaptation

Place plays a defining role in experiencing climate change impacts and how we respond to climatic risks (Devine-Wright, 2013; Hess et al., 2008). This role of place is most apparent through its (1) implications for exposure to climate risks; (2) role in mediating differential vulnerability and adaptive capacity; and (3) outcomes on adaptation decision-making.

The first and most direct relationship between place and climate change is understood through *differential, location-based exposure to climatic risks*. We can intuitively understand that coastal settlements are prone to sea level rise and cyclones, while Arctic communities face more direct impacts of permafrost thawing and associated impacts on hunting and fishing livelihoods. This recognition of place-based risk exposure has led to a focus on defining and mapping 'climate hotspots' (de Sherbinin, 2014; Khan and Cundill, 2019) such as by mapping changes in climate impacts at different levels of warming, often highlighting thresholds and numbers of people impacted (e.g. Lewis et al., 2019). Such a hotspot focus is useful in drawing attention to highly risk-prone locations and populations and providing evidence-based insights into prioritising climate funding and adaptation action (Khan and Cundill, 2019). However, there is less reflection on what gets excluded when we take a hotspot approach—which areas, risks, and people are overlooked[1] and how knowledge creation on certain hotspots might lead to unequal climate action.

The second way place informs climate change research is the recognition that place-based context is a key characteristic of *differential vulnerability and adaptive capacity* (Kelly and Adger, 2000). For example, being closer to district headquarters or having good road infrastructure improves basic public infrastructure and access to essential services (Sewell et al., 2019), both of which shape adaptive capacity to deal with extreme events. Location also determines household access to social and political networks to manage risk (Singh et al., 2018b), abilities to migrate out of risk-prone areas (Sam et al., 2020), and opportunities to diversify livelihoods (Ellis, 2008; Wan et al., 2016). Such an approach has led to creating vulnerability maps where regions, countries, districts, cities, or villages are categorised as having high, medium, or low vulnerability, often based on a range of indicators (Brooks et al., 2005). Such mapping exercises are important but can suffer from an inordinate focus on static representations of vulnerability (Singh et al., 2017) or privilege the importance of location-based exposure in driving vulnerability, over other, socio-cultural and economic factors.

The third way place is critical in climate change research is through its *impacts on adaptation decision-making*, often most visible through place-attachment. Place attachment acknowledges that cultural and emotional ties to a place shapes individual and collective identities and risk perceptions, and informs how people cope with and adapt to risk (Adams and Kay, 2019; Quinn et al., 2018). For example, in high-risk coastal settlements, attachment to fishing livelihoods, place-based cultural practices, and existing social networks inform peoples' decisions to stay rather than relocate even when hazard intensity and frequency increases (Swapan and Sadeque, 2021). This recognition of place as mediating risk perception (and hence adaptation decision-making and action) also elevates the focus on the intangible losses of climate change such as the loss of a way of life (e.g. biodiversity degradation affecting cultural practices) or long-held livelihood practices (e.g. fishing livelihoods eroded due to changing ocean temperatures) (Adger et al., 2013, 2011).

Given this critical role of place in climate change risk exposure, impacts, and adaptation practices, it is not surprising that risk management strategies and plans are deeply rooted in place and highly context-specific. While this centrality of place is deeply acknowledged and well evidenced in climate change research, it has not led to reflexive discussions on how we choose places for climate research and what the implications of such choices are on what/who we identify as vulnerable and what adaptation actions we learn from.

Choosing a place: methodological choices in climate research

In all research, choice of site is a critical decision. It contextualises research questions, shapes approaches to answer them, defines what we find, and reveals personal motivations (Chambers, 1983; Dwyer and Buckle, 2009). Given the urgency of climate action, the inequities it lays bare, and the importance of place in mediating climate impacts and adaptation actions, the consequences of the places we choose to study and experiment in are particularly important.

Currently, risk-centric approaches to choose sites tend to cluster around exposure (e.g. drought-hit villages or flood-prone coastal cities) or ecosystem type (mountains or drylands) (de Sherbinin, 2014), while vulnerability-centric approaches focus on particular groups (e.g. women, certain ethnic groups) (e.g. Rao et al., 2019 examining gender and adaptation across Asia and Africa) or particular trajectories of marginalisation and risk accumulation (e.g. Bankoff, 2003 explains how colonial histories and postcolonial choices shape present-day vulnerability to floods in Manila).

The choice of location in climate change research has repercussions at two scales. Internationally, it has led to a growing critique of identifying solutions that work in particular contexts and applying them to socio-culturally different places without adequate understanding, acknowledgement, and changes made to suit southern risk regimes, contextual risk management practices, and place-specific sociocultural norms and values (Hardoy and Satterthwaite, 1991). This has often led to privileging certain techno-centric or infrastructural adaptation solutions over slower, nature-based, or collective approaches (Nightingale et al., 2020). At a national level, the choice of particular locations has led to empirical evidence bases that draw on the 'usual suspects' of climate vulnerability and adaptation research. For example, when it comes to flood risks, we see a focus on megacities (e.g. studies tend to cluster around Jakarta, Mumbai, and Guangzhou), potentially because of the high absolute damages they are expected to see in monetary terms. This 'metropolitan bias' (Denis and Zérah, 2017) comes at the cost of smaller cities within these countries, obscuring the contextual challenges and strengths of smaller places (Rumbach, 2016).

The consequences of these patterns of choosing certain places have very real repercussions on who is represented or recognised as vulnerable to climate change, which locations are perceived as worthy of investigating climate change impacts and losses and damages, how the drivers and barriers of adaptation action are investigated and understood, and ultimately, what climate solutions are deemed effective and worthy of scaling up (Adger et al., 2003; Fisher, 2015; O'Neill et al., 2010; Popke, 2016). In this sense, choosing a site becomes a value-laden, almost political act; a choice which can become a site of climate justice (whose voice is counted, whose silence is overlooked?); and a forum to diversify knowledge creation (which solutions are we learning from and where are we applying these lessons?).

Overall, place plays a critical role in climate change research but there is relatively limited methodological reflection on how this choice shapes what we look for and what we find. To understand this link between choice of location (for climate change research) and its implications on what we find and whose voices are privileged or silenced, I now discuss two cases from India.

The over- and under-researched in India: illustrative cases

In India, temperatures have already risen by 0.7°C in the twentieth century and are projected to increase by approximately 4.4°C by 2100 (relative to the recent past) (Krishnan et al., 2020). Notably, the impacts of climate change and increasing climate variability are already being experienced, with differential impacts on its farmers, city dwellers, public infrastructure, and livelihoods. This recognition of high climate vulnerability, superimposed on existing development deficits, has led to a mainstreaming of climate action in national and subnational policies. Concurrently, there has been rapid growth in climate change vulnerability and adaptation research across India, especially on highly exposed sectors such as agriculture or in specific at-risk locations such as coastal cities.

I draw on a literature review and personal experiences of over a decade of climate research in India to discuss examples of over- and under-researched rural and urban sites. For the rural sites, in the first section, I reflect on risk management in agrarian livelihoods, highlighting two sites: (1) Mahbubnagar, Telengana in South India, where decades of pioneering field research (1970s–present) have helped develop longitudinal datasets on changing risk and adaptation in drylands; and (2) Pratapgarh, Rajasthan, an agrarian, predominantly tribal district in West India facing high water scarcity but substantially lower research interest. For the former, I draw on the literature while for the latter I use insights from my doctoral research (2010–2014). Mahbubnagar and Pratapgarh are chosen to illustrate over- and under-researched sites, respectively.

For the urban examples, in the second section, I discuss three medium-sized cities—Surat, Indore, and Gorakhpur—to draw attention to how a long-term, internationally funded project led these cities to become iconic examples of urban adaptation in India. Well-documented in peer-reviewed and grey literature, these cities have a rich repository of research and practice-based insights on the risks these cities face and governance challenges and solutions around climate adaptation. I then highlight how this potentially obscures urban adaptation in other cities, drawing on data from a parallel study that reviewed urban adaptation across India (Singh et al., 2021).

Climate change adaptation and rural livelihoods

> *It's rumored, in parts of India, villagers have seen so many researchers that when they see a new one with a questionnaire, they ask if it's for a Master's or a PhD thesis & accordingly make the answers short or long.*
>
> (Basu, 2019)

Village-level studies in Mahbubnagar, Telengana

Mahbubnagar is a semi-arid district in Telengana, a state in South India. Mainly dependent on rainfed agriculture and characterised by small landholdings, the district is emblematic of the 'monsoon-dependent, smallholder farmer' often identified as most vulnerable to climate change in India (Kumar et al., 2016; Wood et al., 2014). Mahbubnagar district has seen sustained research focus and 'action research' projects since the 1970s, with a range of experiments on climate change *impacts* such as on crop germination, fruiting and yields, and diseases as well as climate change *solutions* such as developing drought-tolerant or early maturing crop varieties, examining outcomes of no-till farming, and technological experiments in irrigation, sowing, and post-harvest practices.

Tracing back, this heightened research focus in the district coincides with the setting up of the International Crops Research Institute for the Semi-arid Tropics (ICRISAT) in 1972, a 2–3 hours' drive away from the district headquarters. Based in India, ICRISAT was launched as an international non-profit organisation undertaking agricultural research for development 'through partnerships and with an Inclusive Market Oriented Development' (ICRISAT, 2020).

Sustained funding and a continued research focus on improving agrarian livelihoods enabled an accrual of place-based expertise and detailed, longitudinal data in Mahbubnagar. For example, ICRISAT's longitudinal Village Level Studies (VDSA),[2] started in 1975 (Shambu Prasad et al., 2005), provide a rich, multi-decadal data repository making Jonathan Morduch (2002, p. 38) note, perhaps somewhat hyperbolically, that '*in the history of modern development economics, no single data set has yielded as many important microeconomic papers as the ICRISAT Village Level Studies*'. Seemingly surprised by the richness of this data, Morduch (2002, p. 54) goes on to reflect, '*remarkably, such fine work over the two decades of development economics springs from evidence on just over 100 households. If this is a precedent, creating similar data sets in other regions will surely repay the effort many times over*'. This longitudinal (and importantly, open access) dataset from Mahbubnagar pioneered early work on risk management in dryland areas and continues to inform climate change research in India (e.g. Kumar et al., 2020).

While the products of action research in Mahbubnagar have informed government dryland policy, shaped on-farm experiments and incentives, and been foundational for climate change research in rural India, it is important to recognise that it has been shaped by ICRISAT's techno-centric, market-driven approach, often led by developmental economists and with a bias towards quantified assessments of vulnerability and adaptation decision-making (e.g. Reddy et al., 2010; Shankar et al., 2014; Walker and Ryan, 1991). Further, the VDSA data also face 'attribution, fatigue, and measurement errors' (Morduch, 2002, p. 54), as expected in repeat surveys in particular locations. These limitations of the dataset and how and when lessons from such datasets can be applied to other dryland contexts is less discussed in the literature.

Changing livelihoods in Pratapgarh, Rajasthan

Let us now turn our attention to the other end of the spectrum. In 2011, when I began my doctoral fieldwork on water scarcity and farmer risk management in Rajasthan, Pratapgarh was a newly formed district in the state.[3] My choice was driven by the fact that water scarcity has been over-studied in Rajasthan's arid northern districts with lesser emphasis on the relatively wetter but water-scarce southern districts. Even within South Rajasthan, some districts such as Udaipur, are over-researched and over-implemented in—friends working in local NGOs joked that each village in Udaipur has three NGOs operational—one each for education, environmental, and health issues, and they worked together smoothly, having already 'divided' their sectors and hence spheres of action and influence.

In Pratapgarh, I was confronted by a disconcerting absence of historical data to draw on, the complete lack of a research cohort who knew of the site and its dynamics, and the silence in contemporary climate change studies in both natural and social sciences. The gaps in longitudinal data were a major challenge to constructing timelines of drought and water scarcity for the region. In 2010, when I was preparing for fieldwork, there were no peer-reviewed articles and very little grey literature on Pratapgarh's history of socio-political marginalisation, its peculiar geography of basalt under-rock, its livelihood transformations, and pertinently (for my research), its response to drought and water scarcity. While this paucity of previous data and academic conversation on Pratapgarh was extremely challenging and isolating, it also allowed (or perhaps necessitated?) curiosity and an open approach to data collection. Unbiased by previous findings, I approached the research on perceptions of climate variability in an exploratory manner, slowly constructing narratives of risk and response from colonial district gazetteers, oral histories, repeated visits to neighbouring district libraries, and interviews with local researchers who did not study climate change but had written about tribal issues in south Rajasthan (Singh, 2014a). This approach took more time, often led to dead ends, and sometimes felt like a lot of effort for small results. At times, it was demoralising, as one quote from my fieldnotes recounts:

> Why did you come to Pratapgarh of all places? Whoever told you to come here misguided you terribly. Who would want to come to Pratapgarh when you could go elsewhere? It is very backward and you will have to work with tribals.
>
> District Tribal Development Officer, quoted in Singh 2014a

Some of these challenges were overcome by liaising with a local NGO that was embedded in the region for many years and eased my entry into the research locations. While I did not have the social capital that researchers of over-researched places can draw upon, I built my networks and garnered local interest in my work as I went along (Singh, 2014b).

In retrospect, my choice of Pratapgarh was also an act of dissent, moving away from the usual suspects of dryland rural studies in India, to examine climate risks and responses in a tribal-dominated district that was under-researched. In doing so, I developed a repository of information on a site that is not only experiencing more erratic rain and temperature fluctuations but also seeing socio-political marginalisation. By recording the perceptions and experiences of farmers, traditional healers, and labourers of all genders and castes (Singh et al., 2018a), the work in under-researched Pratapgarh possibly pluralised the knowledge systems drawn upon and voices being heard on climate change perceptions and adaptation decisions in India.

Building climate resilience in urban India

> As a scholarly community, we have invested far more time, energy and resources in understanding the impacts of and adaptation to climate change in some African cases than in others . . . In researching these phenomena, we have been looking disproportionately where the light is most plentiful, rather than where the manifest need or opportunities for affecting outcomes are greatest. Addressing these systematic biases in the state of our knowledge is crucial for both scientific and ethical reasons.
>
> (Hendrix, 2017, p. 146 on the streetlight effect)

While climate change research in India has traditionally focused on rural areas where impacts are more tangible and proximate (e.g. crop failure), there is increasing attention to climate risks and adaptation in Indian cities (Revi, 2008). However, this increasing attention has been concentrated in certain cities based on the international research projects or transnational urban networks they were part of, with less attention to smaller cities, urban settlements in particular geographies, and those that might be at similar risk but no demonstrated climate action (Rumbach, 2016; Rumbach and Follingstad, 2019; Singh et al., 2021). I now discuss examples of over- and under-research in urban India, reflecting on the literature around urban adaptation and resilience building.

Over-researched cities: Surat, Indore, and Gorakhpur

Surat, Indore, and Gorakhpur are often identified as early adaptors in India (Sharma et al., 2014), having benefitted from long-term research and implementation on climate resilience driven by the Rockefeller Foundation–funded Asian Cities Climate Change Resilience Network (ACCCRN) project. Started in 2008, ACCCRN worked with city governments, researchers, and non-governmental organisations to enable evidence-based resilience programming in Indian cities through risk assessments, multi-stakeholder engagement, and sectoral resilience-building interventions (Karanth and Archer, 2014). In 2011–2014, the project expanded to 30 cities focussing on replicating and scaling up of urban climate change resilience actions in India.

Given this sustained funded action, Surat, Gorakhpur, and Indore are regularly highlighted as examples of climate resilience building in India and the Global South. Researchers and practitioners with different foci have used these three cities to build a strong evidence base on resilience conceptualisation and implementation (Bahadur and Tanner, 2014); urban resilience governance (Bellinson and Chu, 2019; Chu, 2016); and the mechanics of mainstreaming climate resilience in urban development (Bhat et al., 2013; Karanth and Archer, 2014; Sharma et al., 2014). Lessons from the cities have been pivotal in inculcating a culture of research-practice partnerships on urban resilience and experimenting with different intervention approaches and institutional arrangements (e.g. the setting up of The Climate Change Trust in Surat or the role of a local NGO in bottom-up action in Gorakhpur).

However, the ACCCRN cities have a disproportionate influence on the discourse on urban climate resilience in India.[4] Collectively, home to about 10.5 million people (2019 estimates), these three cities represent a tiny fraction of India's 470 million strong urban population (UNDESA, 2019). While this does not take away from the rich evidence base the ACCCRN cities provide, it highlights the paucity of other long-term datasets on climate risks and responses in Indian cities. It also potentially renders Surat, Gorakhpur, and Indore as 'totemic examples' of urban resilience in India, making urban resilience discourses suffer from the 'streetlight effect' (Hendrix, 2017). The streetlight effect is described as 'the tendency for researchers to focus on particular questions, cases and variables for reasons of convenience or data availability rather than broader relevance, policy import, or construct validity'. While learning from these cities is not problematic on its own (and might even be encouraged to showcase how different people are converging on similar findings or not), it begs the question, 'what are we not seeing when we are only focussing on only three cities?'

Over-researched 'places' often lead to 'over-researched sectors', further resulting in the peripheralisation of certain sectors, agendas, and people. For example, the literature on climate resilience in Indore tends to focus on water management predominantly, with flood risks mentioned most prominently. In contrast, there has been lower academic commentary on energy, food, or employment aspects shaping Indore's resilience. Even within the water sector, there remain data gaps on groundwater, behavioural change towards demand reduction, and impacts of slow-onset urban droughts. While there is a reason that the over-researched sectors are important (they have after all, over time, been identified as critical issues in the city), they possibly push researchers to continue to study what has been studied because of pre-existing baseline data, ongoing interventions to study, and an established discourse to embed one's arguments into or against. The implications of such peripheralisation of sectors *within* over-researched places is worth examining.

Under-researched places and risks: small towns and slow-onset risks

Recent reviews of urban adaptation research in India points to several gaps, either of particular cities (e.g. small towns and medium-sized cities) or of particular risks (e.g. slow-onset events such as urban droughts and sea level rise) (Singh et al., 2021). There is also a predominance of research on particular income groups, with implications of who is labelled as vulnerable. Let us unpack the implications of under-researched places in urban adaptation in more detail.

First, despite being spaces of rapid change and growing vulnerability, small cities and peri-urban spaces in India are underrepresented in the literature (Rumbach and Follingstad, 2019; Singh and Narain, 2020), often falling through divisions of academic enquiry along rural studies and urban studies. This is most notable in climate change vulnerability assessments where they are negligibly discussed (Singh et al., 2017). Second, there is a discernible focus on particular rapid-onset risks (e.g. flooding in Surat and Gorakhpur) with lesser attention to slow-onset but equally critical risks such as sea level rise and drought (Singh et al., 2021). This unbalanced focus on particular risks (also reflected in the disaster management literature) can undermine the efforts to prepare for multiple risks that occur concurrently and often with interacting, cascading impacts. The case of Chennai, a city facing sea level rise, cycles of drought and localised flooding, and growing groundwater extraction, is a case in point.

Third, certain sectors and social groups within cities are less researched: e.g. urban vulnerability assessments tend to focus on informal settlements and low-income groups. While this focus is critical in understanding how socio-economic marginalisation and climate change vulnerability intersect (e.g. Gajjar et al., 2019), there needs to be a parallel focus on examining the exposure and responsibilities of those with the highest carbon footprints (Bhoyar et al., 2014). Knowledge creation on the poorest and most vulnerable alone can lead to an incomplete mapping of urban vulnerability, mask how actions by certain (elite) groups redistribute risk unevenly, and inadvertently place the burden of adapting on the most vulnerable.

Finally, a focus on few sites as 'totemic examples' for lessons on climate adaptation can possibly obscure other urban places that perhaps need more attention and, in some cases, can provide additional insights on enabling place-based adaptation. Recent work on climate action in Indian cities (Basu and Bazaz, 2018 on Siliguri; Bhardwaj and Khosla, 2020 on Rajkot and Coimbatore) are pluralising the sites studied, the methodologies used, and the lessons for urban resilience but these are yet to balance the over-representation of the over-researched.

Discussion: over-representing and obscuring through over- and under-research

Overall, under- and over-researched places have pros and cons for how we examine climate risks and identify adaptation solutions (Table 10.1). They hold positive and negative implications for what types of examples are drawn upon and learnt from, whose voices are privileged, and which places and people are rendered invisible. I now turn to these concerns.

The most intuitive implication of over-research (in general and in climate change research in particular) is that it leads to *over-representation* of certain conceptual frames, certain places as being more risk-prone, and certain solutions as being most effective. Over-research increases the danger of certain places becoming 'totemic examples' of climate action in urban India and that they may or may not fairly represent the range of climate risks cities face and the variety of institutional and funding arrangements cities are using (Singh et al., 2021).

This over-representation can also lead to certain solutions being seen as 'best practices' and worthy of 'knowledge transfers'. New York and Rotterdam, two

Table 10.1 The pros and cons of over- and under-researched places, especially for climate research

Locations	*Pros*	*Cons*
Over-researched (e.g. Mahbubnagar district, Surat and Gorakhpur, Indore)	– Long-term datasets enable generations of researchers to revisit past data and chart trajectories of change (especially important when looking at long-term changes in climate trends, changing risk perceptions, and changing risk management strategies) – Incremental addition of knowledge, long-term funding allowing for evidence-based implementation – More visibility to highly exposed locations	– Undue focus on particular totemic examples leading to overgeneralisation (e.g. a focus on certain villages might obscure other farming arrangements, cultural contexts, ecologies etc.; certain cities might become shorthand for all high-risk cities) – Researcher bias where past findings influence what one looks for and finds. – Potential methodological conformity due to use of same long-term datasets, similar data gaps. – Respondent fatigue and attrition
Under-researched (Pratapgarh district, peri-urban India, smaller cities)	– Conducive to an open approach to data collection, unbiased by previous findings – Pluralises the places and knowledges being drawn from to construct narratives of climate risk and response – Addresses regional and/or sectoral gaps (e.g. climate change research in small cities or tribal districts)	– Absence of cohort of scholars talking about and publishing research from the location – Lack of long-term datasets to compare findings against – Challenges to have research impact, when evidence base is relatively new

cities that loom large on the research landscape of urban adaptation, are a case in point. Flood management practices from these high-income, industrialised cities have been 'transferred' most notably to high-risk but socio-culturally different cities such as Jakarta and Dhaka. As Thomas (2020, p. 6) notes, from flood risk management in Bangladesh,

> recommending permanent embankments, though, it failed to recognize the difference between the northern temperate climate of The Netherlands and the monsoon-driven climate of South Asia, the difference between essential *borsha* and damaging *bonna* floods, and the existing embankments' seasonal construction and use.

Such a focus on learning from 'best practices' in 'exemplar cities' must be treated with deliberation (Bulkeley, 2006), and if poorly applied, can lead to possibly maladaptive outcomes.

Over-researched places can also constrain inclusive climate research when a few sites homogenise narratives of climate risk perception and climate action (Devine-Wright, 2013). This, I argue, is the biggest bias climate researchers (especially those working on vulnerability and adaptation) must be cognizant of—as we choose our sites and collect our data, we must ask whose voices are privileged and whose are silenced? What bodies of knowledge, what innovations are we failing to learn from as we develop 'best practices' and 'success stories' from a handful of sites? In this sense, the choice of research site becomes political, since it betrays judgements about what places, people, and risks are deemed worthy of research attention. In the current trend towards funding bodies incentivising action research and research-for-impact, this choice of where we examine climate impacts and where we learn from has real-world implications for evidence-based climate action and policymaking.

Finally, over-research not only faces the danger of inaccurately representing the field but also constricts methodological plurality (with implications for the type of data collected and conclusions reached). In Mahbubnagar, this has led to concerns over response fatigue and over-reliance on particular types of quantitative data, while in over-researched Indian cities, this has potentially led to an over-reliance on lessons from few cities, at the cost of other voices, experiences, and lessons (Rumbach, 2016). Methodologically, over-researched places might over carry the weight of 'Ghosts of Researchers Past' (as discussed by Cat Button in her chapter in this volume; Chapter 4) where past researchers and their findings can bias follow-on research through the conceptual frames used to structure the enquiry, the methodological tools and research partners used to enter the field, the analytical devices applied to the data collected, and the results gleaned.

Conclusion

Climate change adaptation has long acknowledged the critical role context plays in differential vulnerability and adaptive capacity. This context is inherently place-based and embedded in histories of ecological, sociocultural, and infrastructural changes (Adger et al., 2011; Devine-Wright, 2013; Ford et al., 2010). In climate research, the role of 'place' is often articulated through ideas of 'climate hotspots' (locations highly exposed to climate risks), location-specific assessments of differential vulnerability and adaptive capacity, and implications of location on adaptation decisions (e.g. place-attachment and its role in people choosing to stay or relocate from hazardous places).

Given this centrality of place in climate change impacts and solutions, I reflect on the implications of choice of research locations when examining climate vulnerability and adaptation. Drawing on illustrative examples of climate change research in rural and urban India, I argue for a more deliberate and reflexive methodological approach to choosing sites of climate research.

Certain places (districts, watersheds, villages, and/or cities) in India are overly represented in the climate literature. These over-researched places are a product of established longitudinal studies and follow funding channels, with research begetting more research. On the positive side, this concentrated attention has led to robust longitudinal datasets as shown in the example of agrarian research in Mahbubnagar. In the urban examples of Surat, Gorakhpur, and Indore, this over-research has shaped the discourse on urban adaptation in India, an increasingly important part of India's climate strategy. However, over-research in certain sites becomes dangerous when lessons and solutions from them are used as totemic examples (Thomas, 2020), leading to the potential (and perhaps unintentional) silencing of other places and their realities and falling back on the 'usual suspects' for identifying solutions.

This brings us to the second reason to reflect on choice of research sites—that of under-researched places. The danger of ignoring under-researched places in climate change research is that we end up telling half-stories, potentially silencing certain experiences and knowledge systems and overlooking critical spaces of challenges and opportunity (Devine-Wright, 2013; Popke, 2016). As I show in the example of Pratapgarh, work in under-researched places can expand our understanding of climate vulnerability and risk management, add to the narratives we construct around climate impacts, and pluralise the suite of solutions to choose from. However, under-researched places are under-researched for a reason, and structural challenges of poor long-term datasets, inadequate understandings about local issues, and relatively less-developed discourses in these areas can perpetuate further under-research. As attention clusters around over-researched sites, these under-researched places (also exposed to climate risks and sometimes, experimenting with solutions) repeatedly fall off the map.

Recognising the power of place in climate research is critical to respond to growing calls of inclusive climate research that captures the experiences of different places and people and develops evidence of experiments and solutions that are place-based. There is of course no one 'right' place to conduct research; however, recognising the biases and benefits that a location signifies is critical for more deliberative research to not only build upon existing work but also to give voice to silences in places that are under-researched. As reflexive climate researchers tuned into discourses of climate justice and inclusive adaptation, a critical focus on the choice of location begins with a simple question—what are we possibly not seeing, counting, or acknowledging, when we focus on *this* place?

Notes

1 In his review of climate hotspots research, de Sherbinin (2014) finds that mapping exercises tend to focus on identifying high-risk places based on biophysical vulnerability with lesser attention to social vulnerability.
2 The Village Level Studies (VLS) are now also called the Village Dynamics Studies in South Asia (VDSA). For details, see http://vdsa.icrisat.ac.in/.
3 Pratapgarh was carved out of the historically and culturally significant Chittorgarh district in January 2008, about 3.5 years before I visited for a year-long fieldwork (September 2011–August 2012).
4 From a list of 213 articles on climate adaptation in India, 66 (31%) use examples from the ACCCRN project. This list was from a search on Scopus conducted on October 20, 2020, using the search string ALL (*"clim* change"* OR *"clim* variability"*) AND TITLE-ABS-KEY (*India*) AND TITLE-ABS-KEY (*adapt** OR *resilien**) AND TITLE-ABS-KEY (*"city"* OR *"urban"* OR *"town"*), which gave 322 hits. An abstract scan led to further reductions by removing inapplicable articles (e.g. articles from medical sciences on evolutionary adaptation or those that do not explicitly report on urban adaptation or resilience in India) and ended with a list of 213 papers. Of these, 66 focused on ACCCRN cities.

References

Adams, H., Kay, S. (2019) Migration as a human affair: Integrating individual stress thresholds into quantitative models of climate migration. *Environmental Science and Policy* 93, 129–138. https://doi.org/10.1016/j.envsci.2018.10.015

Adger, N., Barnett, J., Brown, K., Marshall, N., O'Brien, K. (2013) Cultural dimensions of climate change impacts and adaptation. *Nature Climate Change* 3, 112–117. https://doi.org/10.1038/nclimate1666

Adger, N., Huq, S., Brown, K., Conway, D., Hulme, M. (2003) Adaptation To Climate Change in the Developing World. *Progress in Development Studies* 3, 179–195. https://doi.org/10.1191/1464993403ps060oa

Adger, W.N., Barnett, J., Chapin, F.S., Ellemor, H. (2011) This Must Be the Place: Underrepresentation of Identity and Meaning in Climate Change Decision-Making. *Global Environmental Politics* 11, 1–25. https://doi.org/10.1162/GLEP_a_00051

Bahadur, A. V., Tanner, T. (2014) Policy climates and climate policies: Analysing the politics of building urban climate change resilience. *Urban Climate* 7, 20–32. https://doi.org/10.1016/j.uclim.2013.08.004

Bankoff, G. (2003) Constructing vulnerability: The historical, natural and social generation of flooding in metropolitan Manila. *Disasters* 27, 224–238. https://doi.org/10.1111/1467-7717.00230

Basu, K. (2019) Economic Graffiti: The view from Palanpur. *Indian Express*. 15 June 2019. https://indianexpress.com/article/opinion/columns/the-view-from-palanpur-christopher-bliss-nicholas-stern-5779621/

Basu, R., Bazaz, A.B. (2018) Reimagining development practice: Mainstreaming justice into planning frameworks, in: Jafry, T. (Ed.), *Routledge Handbook of Climate Justice*. Routledge, London. https://doi.org/10.4324/9781315537689

Bellinson, R., Chu, E. (2019) Learning pathways and the governance of innovations in urban climate change resilience and adaptation. *Journal of Environmental Policy & Planning* 21, 76–89. https://doi.org/10.1080/1523908X.2018.1493916

Bhardwaj, A., Khosla, R. (2020) Superimposition: How Indian city bureaucracies are responding to climate change. *Environment and Planning E: Nature and Space* 0, 251484862094909. https://doi.org/10.1177/2514848620949096

Bhat, G.K., Karanth, A., Dashora, L., Rajasekar, U. (2013) Addressing flooding in the city of Surat beyond its boundaries. *Environment & Urbanisation* 25, 429–441. https://doi.org/10.1177/0956247813495002

Bhoyar, S.P., Dusad, S., Shrivastava, R., Mishra, S., Gupta, N., Rao, A.B. (2014) Understanding the Impact of Lifestyle on Individual Carbon-footprint. *Procedia—Social and Behavioral Sciences* 133, 47–60. https://doi.org/10.1016/j.sbspro.2014.04.168

Brooks, N., Adger, W.N., Kelly, P.M. (2005) The determinants of vulnerability and adaptive capacity at the national level and the implications for adaptation. *Global Environmental Change* 15, 151–163. https://doi.org/10.1016/j.gloenvcha.2004.12.006

Bulkeley, H. (2006) Urban sustainability: Learning from best practice? *Environment and Planning A: Economy and Space* A 38, 1029–1044. https://doi.org/10.1068/a37300

Chambers, R. (1983) *Rural development: Putting the last first*. Prentice Hall, Harlow.

Chu, E. (2016) The political economy of urban climate adaptation and development planning in Surat, India. *Environment and Planning C: Government and Policy* 34, 281–298.

de Sherbinin, A. (2014) Climate change hotspots mapping: What have we learned? *Climatic Change* 123, 23–37. https://doi.org/10.1007/s10584-013-0900-7

Denis, E., Zérah, M.-H. (2017) *Subaltern urbanisation in India: An introduction to the dynamics of ordinary towns*. Springer, New Delhi. https://doi.org/10.1007/978-81-322-3616-0

Devine-Wright, P. (2013) Think global, act local? The relevance of place attachments and place identities in a climate changed world. *Global Environmental Change* 23, 61–69. https://doi.org/10.1016/j.gloenvcha.2012.08.003

Dwyer, S.C., Buckle, J.L. (2009) The space between: On being an insider-outsider in qualitative research. *International Journal of Qualitative Methods* 8, 54–63. https://doi.org/10.1177/160940690900800105

Ellis, F. (2008) The determinants of rural livelihood diversification in developing countries. *Journal of Agricultural Economics* 51, 289–302. https://doi.org/10.1111/j.1477-9552.2000.tb01229.x

Fawcett, D., Pearce, T., Ford, J.D., Archer, L. (2017) Operationalizing longitudinal approaches to climate change vulnerability assessment. *Global Environmental Change* 45, 79–88. https://doi.org/10.1016/j.gloenvcha.2017.05.002

Fisher, S. (2015) The emerging geographies of climate justice. *The Geographical Journal* 181, 73–82. https://doi.org/10.1111/geoj.12078

Ford, J.D., Keskitalo, E.C.H., Smith, T., Pearce, T., Berrang-Ford, L., Duerden, F., Smit, B. (2010) Case study and analogue methodologies in climate change vulnerability research. *Wiley Interdisciplinary Reviews: Climate Change* 1, 374–392. https://doi.org/10.1002/wcc.48

Gajjar, S.P., Jain, G., Michael, K., Singh, C. (2019) Entrenched vulnerabilities: Evaluating climate justice across development and adaptation responses in South India, in: *Climate futures: Reimagining global climate justice*. Zed Books, London.

Hardoy, J.E., Satterthwaite, D. (1991) Environmental problems of third world cities: A global issue ignored? *Public Administration and Development* 11, 341–361. https://doi.org/10.1002/pad.4230110405

Hendrix, C.S. (2017) The streetlight effect in climate change research on Africa. *Global Environmental Change* 43, 137–147. https://doi.org/10.1016/j.gloenvcha.2017.01.009

Hess, J.J., Malilay, J.N., Parkinson, A.J. (2008) Climate change. The importance of place. *American Journal of Preventive Medicine* 35, 468–478. https://doi.org/10.1016/j.amepre.2008.08.024

ICRISAT (2020) *Strategic plan to 2020: Inclusive market-oriented development for smallholder farmers in the tropical drylands* [WWW Document]. URL www.icrisat.org/newsroom/latest-news/one-pager/sp2020/sp-2020.htm (accessed 1.18.21).

Karanth, A., Archer, D. (2014) Institutionalising mechanisms for building urban climate resilience: Experiences from India. *Development in Practice* 24, 514–526. https://doi.org/10.1080/09614524.2014.911246

Kelly, P.M., Adger, W.N. (2000) Theory and practice in assessing vulnerability to climate change and facilitating adaptation. *Climatic Change* 47, 325–352.

Khan, A.S., Cundill, G. (2019) Hotspots 2.0: Toward an integrated understanding of stressors and response options. *Ambio* 48, 639–648. https://doi.org/10.1007/s13280-018-1120-1

Krishnan, R., Sanjay, J., Gnanaseelan, C., Mujumdar, M., Chakraborty, Ashwini Kulkarni, S. (2020) *Assessment of climate change over the Indian region*. Springer Singapore, Singapore. https://doi.org/10.1007/978-981-15-4327-2

Kumar, S., Mishra, A.K., Pramanik, S., Mamidanna, S., Whitbread, A. (2020) Climate risk, vulnerability and resilience: Supporting livelihood of smallholders in semiarid India. *Land Use Policy* 97, 104729. https://doi.org/10.1016/j.landusepol.2020.104729

Kumar, S., Raizada, A., Biswas, H., Srinivas, S., Mondal, B. (2016) Application of indicators for identifying climate change vulnerable areas in semi-arid regions of India. *Ecological Indicators* 70, 507–517. https://doi.org/10.1016/j.ecolind.2016.06.041

Lewis, S.C., King, A.D., Perkins-Kirkpatrick, S.E., Mitchell, D.M. (2019) Regional hotspots of temperature extremes under 1.5°C and 2°C of global mean warming. *Weather and Climate Extremes* 26, 100233. https://doi.org/10.1016/j.wace.2019.100233

Morduch, J. (2002) *Consumption smoothing across space: Testing theories of risk-sharing in the ICRISAT study region of south India*. WIDER Working Paper Series DP2002-55, World Institute for Development Economic Research (UNU-WIDER).

Nightingale, A.J., Eriksen, S., Taylor, M., Forsyth, T., Pelling, M., Newsham, A., Boyd, E., Brown, K., Harvey, B., Jones, L., Bezner Kerr, R., Mehta, L., Naess, L.O., Ockwell, D., Scoones, I., Tanner, T., Whitfield, S. (2020) Beyond technical fixes: Climate solutions and the great derangement. *Climate & Development* 12(4), 343–352. https://doi.org/10.1080/17565529.2019.1624495

O'Neill, S.J., Hulme, M., Turnpenny, J., Screen, J.A. (2010) Disciplines, geography, and gender in the framing of climate change. *Bulletin of the American Meteorological Society* 91, 997–1002. https://doi.org/10.1175/2010BAMS2973.1

Popke, J. (2016) Researching the hybrid geographies of climate change: Reflections from the field. *Area* 48, 2–6. https://doi.org/10.1111/area.12220

Quinn, T., Bousquet, F., Guerbois, C., Sougrati, E., Tabutaud, M. (2018) The dynamic relationship between sense of place and risk perception in landscapes of mobility. *Ecology & Society* 23(2), 39. https://doi.org/10.5751/ES-10004-230239

Rao, N., Mishra, A., Prakash, A., Singh, C., Qaisrani, A., Poonacha, P., Vincent, K., Bedelian, C. (2019) A qualitative comparative analysis of women's agency and adaptive capacity in climate change hotspots in Asia and Africa. *Nature Climate Change* 9, 964–971. https://doi.org/10.1038/s41558-019-0638-y

Reddy, V.R., Brown, P., Bandi, M., Chiranjeevi, T., Reddy, D.R., Roth, C. (2010) *Adapting to climate variability in semi-arid regions: A study using sustainable rural livelihoods framework*. LNRMI Working Paper No. 1.

Revi, A. (2008) Climate change risk: An adaptation and mitigation agenda for Indian cities. *Environment & Urbanisation* 20, 207–229. https://doi.org/10.1177/0956247808089157

Rumbach, A. (2016) Disaster governance in small urban places: Issues, trends, and concerns, in: *Disaster governance in urbanising Asia*. Springer Singapore, Singapore, pp. 109–125. https://doi.org/10.1007/978-981-287-649-2_6

Rumbach, A., Follingstad, G. (2019) Urban disasters beyond the city: Environmental risk in India's fast-growing towns and villages. *International Journal of Disaster Risk Reduction* 34, 94–107. https://doi.org/10.1016/j.ijdrr.2018.11.008

Sam, A.S., Padmaja, S.S., Kächele, H., Kumar, R., Müller, K. (2020) Climate change, drought and rural communities: Understanding people's perceptions and adaptations in rural eastern India. *International Journal of Disaster Risk Reduction* 44, 101436. https://doi.org/10.1016/j.ijdrr.2019.101436

Sewell, S.J., Desai, S.A., Mutsaa, E., Lottering, R.T. (2019) A comparative study of community perceptions regarding the role of roads as a poverty alleviation strategy in rural areas. *Journal of Rural Studies* 71, 73–84. https://doi.org/10.1016/j.jrurstud.2019.09.001

Shambu Prasad, C., Hall, A., Wani, S. (2005) *Institutional history of watershed research: The evolution of ICRISAT's work on natural resources in India*. Global Theme on Agroecosystems Report no. 12, ICRISAT, Patancheru, India.

Shankar, K.R., Nagasree, K., Nirmala, G., Prasad, M.S., Venkateswarlu, B., Srinivasa Rao, C. (2014) Climate change and agricultural adaptation in South Asia, in: *Handbook of climate change adaptation*. Springer Berlin Heidelberg, Berlin, Heidelberg, pp. 1–13. https://doi.org/10.1007/978-3-642-40455-9_50-1

Sharma, D., Singh, Raina, Singh, Rozita (2014) Building urban climate resilience: Learning from the ACCCRN experience in India. *International Journal of Urban Sustainable Development* 6, 133–153. https://doi.org/10.1080/19463138.2014.937720

Singh, A.K., Narain, V. (2020) Lost in transition: Perspectives, processes and transformations in Periurbanizing India. *Cities* 97, 102494. https://doi.org/10.1016/j.cities.2019.102494

Singh, C. (2014a) Understanding water scarcity and climate variability: An exploration of farmer vulnerability and response strategies in northwest India. Unpublished PhD Thesis, University of Reading. Reading, United Kingdom.

Singh, C. (2014b) Researcher's social capital: Liaising with local actors for effective ethnographic research [WWW Document]. *LSE Field Research Method Lab Blog*. URL https://blogs.lse.ac.uk/fieldresearch/2014/06/12/researchers-social-capital-liaising-with-local-actors/ (accessed 10.19.20).

Singh, C., Deshpande, T., Basu, R. (2017) How do we assess vulnerability to climate change in India? A systematic review of literature. *Regional Environmental Change* 17, 527–538. https://doi.org/10.1007/s10113-016-1043-y

Singh, C., Madhavan, M., Arvind, J., Bazaz, A. (2021) Climate change adaptation in Indian cities: A review of existing actions and spaces for triple wins. *Urban Climate* 36, 100783. https://doi.org/10.1016/j.uclim.2021.100783

Singh, C., Osbahr, H., Dorward, P. (2018a) The implications of rural perceptions of water scarcity on differential adaptation behaviour in Rajasthan, India. *Regional Environmental Change* 18, 2417–2432. https://doi.org/10.1007/s10113-018-1358-y

Singh, C., Rahman, A., Srinivas, A., Bazaz, A. (2018b) Risks and responses in rural India: Implications for local climate change adaptation action. *Climate Risk Management* 21, 52–68. https://doi.org/10.1016/j.crm.2018.06.001

Swapan, M.S.H., Sadeque, S. (2021) Place attachment in natural hazard-prone areas and decision to relocate: Research review and agenda for developing countries. *International Journal of Disaster Risk Reduction* 52, 101937. https://doi.org/10.1016/j.ijdrr.2020.101937

Thomas, K.A. (2020) The problem with solutions: Development failures in Bangladesh and the interests They obscure. *Annals of the American Association of Geographers* 110(5), 1631–1651. https://doi.org/10.1080/24694452.2019.1707641

UNDESA (2019) *World urbanization prospects: The 2018 revision (ST/ESA/SER.A/420)*. New York: United Nations Department of Economic and Social Affairs (UN DESA), United Nations.

Walker, T.S., Ryan, J.G., 1991. Village and household economies in India's semi-arid tropics, *Agricultural Systems*. https://doi.org/10.1016/0308-521X(91)90079-P

Wan, J., Li, R., Wang, W., Liu, Z., Chen, B., 2016. Income diversification: A strategy for rural region risk management. *Sustainability* 8, 1064. https://doi.org/10.3390/su8101064

Wood, S.A., Jina, A.S., Jain, M., Kristjanson, P., DeFries, R.S. (2014) Smallholder farmer cropping decisions related to climate variability across multiple regions. *Global Environmental Change* 25, 163–172. https://doi.org/10.1016/j.gloenvcha.2013.12.011

Index